SPECT IMAGING OF THE BRAIN

The colour section (pages 179–188) has been made possible by an unrestricted educational grant from Janssen-Cilag Ltd.

JANSSEN-CILAG Ltd

Developments in Nuclear Medicine

VOLUME 29

Series Editor: Peter H. Cox

The titles published in this series are listed at the end of this volume.

SPECT Imaging of the Brain

edited by

RODERICK DUNCAN
Department of Neurology, Institute of Neurological Sciences,
Southern General Hospital, Glasgow, Scotland

KLUWER ACADEMIC PUBLISHERS
DORDRECHT / BOSTON / LONDON

Library of Congress Cataloging-in-Publication Data

SPECT imaging of the brain / edited by R. Duncan.
p. cm. -- (Developments in nuclear medicine : v. 29)
Includes index.
ISBN 0-7923-4186-4 (alk. paper)
1. Brain--Tomography. I. Duncan, R. II. Series.
[DNLM: 1. Brain Diseases--radionuclide imaging. 2. Mental Disorders--radionuclide imaging. 3. Tomography, Emission-Computed, Single-Photon. W1 DE998KF v. 29 1996 / WL 348 S741 1996]
RC386.6.T65S67 1996
616.8'047572--dc20
DNLM/DLC
for Library of Congress 96-28961

Distributors

for the United States and Canada: Kluwer Academic Publishers, PO Box 358, Accord Station, Hingham, MA 02018-0358, USA
for all other countries: Kluwer Academic Publishers Group, Distribution Center, PO Box 322, 3300 AH Dordrecht, The Netherlands

A catalogue record for this book is available from the British Library

ISBN 0-7923-4186-4

Published in the United Kingdom by Kluwer Academic Publishers, PO Box 55, Lancaster, UK.

Kluwer Academic Publishers BV incorporates the publishing programmes of D. Reidel, Martinus Nijhoff, Dr W. Junk and MTP Press.

Typeset by Speedlith Photo Litho Ltd., Stretford, Manchester, UK.
Printed and bound in Great Britain by Hartnolls Ltd., Bodmin, Cornwall.

Table of contents

List of contributors

IAN BONE
Department of Neurology
Institute of Neurological Sciences
Southern General Hospital
Govan Road
Glasgow G51 4TF
Scotland

GARTH S. CRUICKSHANK
Department of Neurosurgery
Institute of Neurological Sciences
Southern General Hospital
Govan Road
Glasgow G51 4TF
Scotland

RODERICK DUNCAN
Department of Neurology
Institute of Neurological Sciences
Southern General Hospital
Govan Road
Glasgow G51 4TF
Scotland

DONALD G. GROSSET
Department of Neurology
Institute of Neurological Sciences
Southern General Hospital
Govan Road
Glasgow G51 4TF
Scotland

PETER MATHEW
Department of Neurosurgery
Institute of Neurological Sciences
Southern General Hospital
Govan Road
Glasgow G51 4TF
Scotland

JAMES PATTERSON
Department of Clinical Physics
Institute of Neurological Sciences
Southern General Hospital, NHS Trust
Govan Road
Glasgow G51 4TF
Scotland

MARTIN TURNER
Larkfield Centre
Garngaber Avenue
Lenzie G66 3UG
Scotland

J.T. LINDSAY WILSON
Department of Psychology
University of Stirling
Stirling FK9 4LA
Scotland

DAVID J. WYPER
Department of Clinical Physics
Institute of Neurological Sciences
Southern General Hospital, NHS Trust
Govan Road
Glasgow G51 4TF
Scotland

Introduction

In the developed world, images of brain structure are available as an everyday diagnostic aid, and the characteristic appearances of most pathological conditions can be looked up in a textbook. Functional brain imaging is to this day less widely used, partly because most pressing diagnostic questions can be answered by reference to the patient's cerebral anatomy, partly for reasons of technical limitations of functional techniques. PET as a technique is sufficiently resource-demanding and complex to inhibit its use as an everyday diagnostic technique. SPECT lacked suitable tracers for many years, and early systems had poor spatial resolution.

However, rotating gamma camera technology has advanced to the point where images of the brain of reasonable quality can be obtained at most large hospitals, and practical tracers, particularly of regional cerebral blood flow, are easily available. As research advances, clinical applications are emerging. A recent report of the Therapeutics and Technology Assessment Subcommittee of the American Academy of Neurology[1] details a number of currently recognised clinical applications, some of which are dealt with in this book. Given this recognition, it is increasingly important that clinicians (particularly neuroclinicians, psychiatrists and specialists in cerebrovascular disease), nuclear medicine specialists and physicists acquire an idea of the major applications of the technique, and the research background on which these applications are based.

This book does not pretend to cover all applications of SPECT. It confines itself rather to major pathological areas, i.e. epilepsy, cerebrovascular disease, cerebral malignancy, head injury and psychiatric illness, aiming to give the reader an overview of clinical and research applications in each. Where practical guidance is appropriate (e.g. ictal SPECT in epilepsy), this is also given. The book starts with a technical chapter aimed particularly at clinicians; a basic understanding of how a technique works allows an appreciation of its strengths and weaknesses, and thereby a better understanding of the results.

References

1. Assessment of brain SPECT. Report of the Therapeutics and Technology Assessment Subcommittee of the American Academy of Neurology. Neurology, 1996;46:278–285.

R. Duncan (ed): SPECT Imaging of the Brain, ix.

1. Basics of SPECT

JAMES PATTERSON and DAVID J. WYPER

Introduction

To make the best use of any technique, it is important to have a clear idea of the strengths, weaknesses and limitations of the data it produces. This requires an understanding of the technical and scientific bases of the technique. It is assumed that specialists in nuclear medicine will already have this understanding, and this chapter is largely directed toward clinicians intending to use SPECT for clinical or research purposes.

The development of emission tomography is a good example of the fusion of a number of scientific and medical disciplines to produce an effective imaging technique. Each image is the end result of the physical production of a radionuclide, the labelling of that nuclide to a chemical tracer, the administration of the resulting ligand to a patient, the detection of the emitted radioactivity using a scanner and the reconstruction of the information from the scanner's detectors to reproduce the distribution of the radionuclide in graphic form. There are two different techniques of emission tomography: positron emission tomography (PET), is based on radionuclides which decay by positron emission, while single photon emission computed tomography (SPECT, or sometimes SPET) is based on radionuclides which emit gamma rays or X-rays. While PET has some inherent technical advantages over SPECT, economic reality dictates that SPECT is usually the only technique available in routine clinical practice. Recent innovations in the design of multi-head SPECT systems, which allow them to detect positron-emitting radionuclides, have diminished the sharp distinction between the two techniques.

Radioactivity

There are more than 100 atomic elements, each made up of a positively charged nucleus surrounded by a negatively charged 'cloud' of electrons. The fundamental factor which distinguishes one element from another is the number of protons within the nucleus: this is referred to as the atomic number (Z). The number of protons in an atom is balanced by the number of negatively charged electrons, making the atom electrically neutral. If an electron is removed or added, the atom is said to be ionized and will have a resulting charge. The number of electrons in a non-ionized atom determines its chemical behaviour.

R. Duncan (ed): SPECT Imaging of the Brain, 1–42.

The nucleus also contains neutrons, which are uncharged particles with approximately the same mass as protons. The number of protons plus the number of neutrons makes up the atomic weight (A). A particular nucleus with a specific number of protons and neutrons is known as a nuclide and is denoted by the symbol $^{Z}_{A}X$, where X is the chemical symbol of the element. There are many more nuclides (approximately 1700) than elements since each element can have different numbers of neutrons. The different nuclides of an element are referred to as isotopes: isotopes of any one element must have the same number of protons, and hence the same atomic number, but have different numbers of neutrons, hence different atomic masses, e.g. $^{11}_{6}C$, $^{12}_{6}C$, $^{13}_{6}C$, $^{14}_{6}C$. Isotopes can also differ in their nuclear energy states (the same numbers of protons and neutrons are in a different configuration within the nucleus). They are then classed as isomers, e.g. $^{99}_{43}TC$, $^{99m}_{43}Tc$. Most of the nuclei found in nature are stable and retain the same structure indefinitely. The vast majority of known nuclei, however, are unstable and undergo a transformation to a more stable form. This process of radioactive decay alters the mass and/or energy of the nucleus and occurs over a period ranging from a fraction of a second to millions of years. About 1400 radioactive nuclides are known, each of which has a unique and unalterable decay time (see below).

The change from an unstable to a stable configuration is accompanied by the emission from the nucleus of nuclear particles or electromagnetic radiation. Alpha particles, beta particles and gamma rays are the major forms of radioactive emission from the nucleus. Secondary processes within the electron shells result in the emission of X-rays and electrons. Although X-rays are physically indistinguishable from gamma rays they are given this name to differentiate their origin.

An alpha particle consists of two protons and two neutrons bound together. It therefore has an electrical charge of plus two. Beta particles come in two forms ($\beta-$ and $\beta+$) and result from the transformation of a neutron to a proton and vice versa. The $\beta-$ emission has a single negative charge and is equivalent to emission of an electron whereas the $\beta+$ emission has a single positive charge and is equivalent to a positively charged electron, or positron. After emission, positrons themselves take part in a secondary process which has great significance to imaging. As a positron slows down and encounters an electron the two particles undergo an annihilation reaction where the mass of both particles disappears and is replaced by two gamma rays of equal energy (511 keV) travelling in opposite directions. Detection of these two coincident photons at 180° to each other forms the basis of positron emission tomography.

Smaller amounts of energy can be emitted as gamma rays, a form of electromagnetic energy and part of the electromagnetic spectrum (Figure 1.1). Unlike light rays and radio waves, gamma rays and X-rays have enough energy to remove an electron from one of the electron shells in an atom and are referred to as ionizing radiation. A gamma ray emitted by ^{99m}Tc, for example, is 100 000 times more powerful than a photon in the visible part of the spectrum.

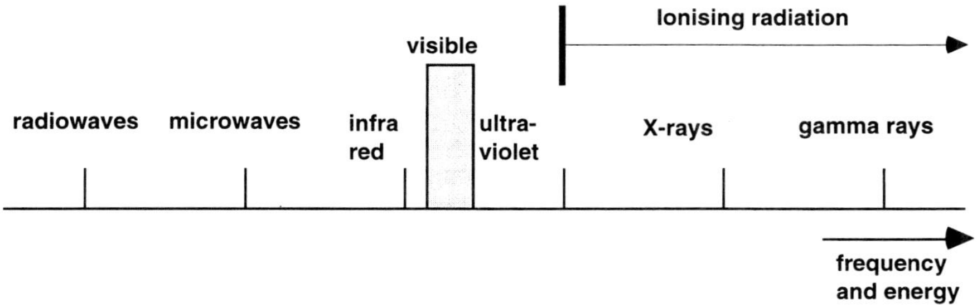

Figure 1.1. Spectrum of electromagnetic radiation.

Quantities and units used for radioactivity

Until recently, the standard unit of radioactivity was the curie (Ci) which was defined, for historical reasons, as the amount of radioactive material in which there are 3.7×10^{10} radioactive disintegrations per second. This is a very large unit and in medicine activity is usually quoted in millicuries (mCi). The unit now more commonly used to measure the quantity of radioactivity is the rather more straightforward becquerel (Bq), which is equal to one disintegration per second. This, by comparison, is a very small unit and the amount of radioactivity administered to a patient for imaging is usually measured in megabecquerels ($1\,\text{MBq} = 10^6\,\text{Bq}$).

The energy of all radiations and particles is expressed in electron volts (eV), a term used throughout nuclear physics. The minimum energy required to remove an electron from the outer shell of an atom, and hence cause ionization, is about 1 eV. The photons emitted by the radionuclides used in medical imaging, with energies five orders of magnitude higher, can dislodge electrons even from the innermost shells of an atom.

Physical and biological half-life

Like many natural phenomena, radioactive decay is an entirely random process and for a given radionuclide some of the nuclei will decay after a very short time while others will disintegrate only after a protracted period of time. Despite this, radioactive decay follows a simple pattern whereby half of the activity remaining at any time subsequently disappears over a period of time known as the physical half-life of the radionuclide. In successive half-lives the activity is reduced to 1/2, 1/4, 1/8, 1/16 and so on. Mathematically, the decay curve is described as being exponential (Figure 1.2).

It is not only the physical decay of a radionuclide which is important in imaging. Once a radionuclide has been incorporated into a larger molecule, it is this molecule which determines its distribution and persistence in the body. In addition to the physical half-life ($T_{½p}$) the radioactivity is cleared from the body by the natural biological washout of the chemical compound of which the radionuclide is part.

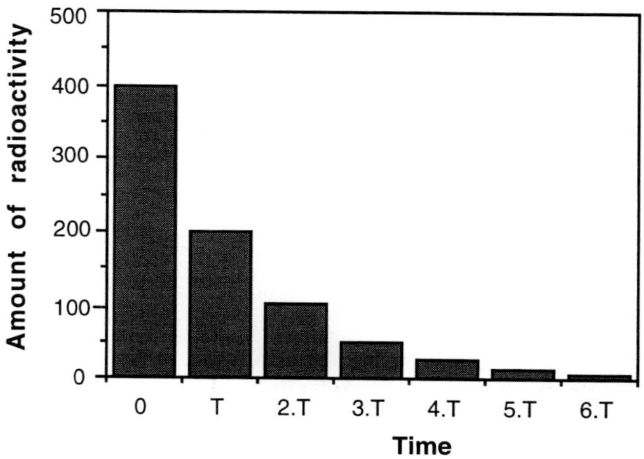

Figure 1.2. Radioactive decay.

This process is usually characterized by an exponential clearance time, resulting in a biological half life ($T_{½b}$). The effective clearance of radioactivity from the body ($T_{½e}$) is therefore a combination of these two processes and is computed from:

$$\frac{1}{T_{½e}} = \frac{1}{T_{½p}} + \frac{1}{T_{½b}}$$

Radionuclides used in medical imaging

Of the 1400 or so known radionuclides very few have the particular blend of characteristics required to make them suitable for use in medical imaging. An appropriate radionuclide must:

1. Emit photons (X or γ rays, or positrons followed by γ rays). It is self-evident that imaging demands the use of radiation which can pass easily through the body so that a radionuclide lying deep within an organ can be 'seen' by an external detector. Only photons have this ability. Alpha particles travel virtually no distance within tissues and beta particles are readily absorbed by a few millimetres of tissue.

2. Emit photons with a suitable energy. For SPECT, the ideal gamma ray energy probably lies between 100 keV and 200 keV, although radionuclides with energies just above and below these values are commonly used. Two fundamental physical problems have a much greater effect below 100 keV; absorption, which reduces the number of gamma rays which can escape from the body to be detected, especially from deep regions, and scattering which

degrades the spatial resolution of the technique. Above 200 keV detector performance is degraded by the ability of these higher energy gamma rays to penetrate the collimator septa (the collimator is the part of the SPECT camera which provides spatial localization of the signal, see Figure 1.6). Spatial resolution is, therefore, impaired at high energies.

3. Emit no alpha particles and as few beta particles or electrons as possible. Any radiation emitted by a radionuclide which does not contribute to the image will deliver an unnecessary radiation dose to the patient. Both alpha and beta particles produce high, localized radiation doses.

4. Have a half-life within a suitable range. The ideal half-life represents a balance between a number of factors. It should be short enough to ensure that the radionuclide does not continue to deliver an unnecessary radiation dose to the patient long after the investigation is complete. On the other hand it must be long enough to allow delivery to the department, labelling of the pharmaceutical, administration to the patient and scanning to be carried out. For the very short half-life nuclides used in PET an expensive on-site cyclotron and highly specialized pharmaceutical facilities are required to fulfil these conditions. The physical and biological half-lives of the radiopharmaceutical also must be long enough to allow optimal imaging of the process under investigation. For some tracers or physiological processes this may require radionuclides with half-lives of many hours or even several days to ensure that high specific binding (e.g. to neuroreceptors) coupled with the clearance of non-specific binding has been achieved. Imaging as late as 24 h post-injection has been used for SPECT, an advantage with respect to PET[1,2].

5. Be able to be labelled to suitable tracers. The final requirement is that the radionuclide can be tagged to an appropriate molecule to produce the desired radioactive tracer (or radiopharmaceutical). For some radionuclides this can be relatively straightforward but, especially for technetium-99m, the chemistry is often complex and demands significant resources for tracer development. PET chemistry, although made easier by the biological nature of the radionuclides available, is also demanding because short half-lives limit the time available for completing the chemical syntheses.

The physical half-lives and the principal emissions for the radionuclides most commonly used in medical imaging are shown in Table 1.1. The first five are used in SPECT while the others can be detected only by a dedicated PET camera or a specially designed SPECT system. Most radiopharmaceuticals used for SPECT imaging have an effective half-life of a few hours.

^{99m}Tc is the most attractive of the SPECT radionuclides. It is available at an affordable cost, at any time, in any nuclear medicine department. It is the daughter product of the longer lived ^{99}Mo ($T_{1/2}$ = 2.8 days) which is supplied to the department in a 'generator' system each week. The ^{99m}Tc is separated each day from

Table 1.1. Physical half-lives and photon energies of radionuclides.

Radionuclide	Principal emissions	Half-life	Photon energy (keV)
^{99m}Tc	γ	6.0 h	140
^{123}I	γ	13.2 h	159
^{111}In	γ	2.83 days	171 and 245
^{201}Tl	X	3.05 days	68–82
^{133}Xe	$\beta+\gamma$	5.25 days	81
^{15}O	Positron (γ)	122 s	511
^{13}N	Positron (γ)	10.0 min	511
^{11}C	Positron (γ)	20.4 min	511
^{18}F	Positron (γ)	109.8 min	511

^{99}Mo by passing sterile saline through the generator column, forming soluble sodium pertechnetate.

The other SPECT radionuclides are produced in either reactors or cyclotrons, and tend to be more expensive. They are available only on particular days of the week but if their overall use is high enough it becomes economical and practical to hold a reasonable quantity as stock for use when required. ^{123}I, because of its shorter half-life, generally has to be used within a day of delivery, making greater demands on patient scheduling, and on chemical synthesis when appropriate.

Radiopharmaceuticals

Functional imaging can be performed using radionuclides in their most basic forms. For example, ^{133}Xe gas in its elemental form can be used to measure cerebral blood flow and assess lung ventilation. The chloride salt of ^{201}Tl is widely used to investigate myocardial perfusion and to assist in the evaluation of brain tumours. These are, however, exceptions, and the future of functional imaging relies on the development of molecules specifically designed to reflect the physiological biochemistry of either an organ, a type of cell, a cellular process or a receptor system[3]. There are many approaches to the development of radiopharmaceuticals, physical, chemical, physiological and pharmacological research each playing their part[4]. Molecular biology, computer-aided design and genetic engineering may also make major contributions and it would be surprising if serendipity did not have the same impact as it has in conventional drug design.

Unlike PET, where direct substitution is possible with carbon-11, the incorporation of any of the non-ideal SPECT nuclides into a molecule alters the behaviour of that molecule and makes rational design difficult[3]. While iodine can be substituted fairly easily (e.g. for a hydroxyl group) it increases lipid solubility, leading to higher non-specific binding and more flow dependency during early uptake. Steric effects, such as have been found with ^{123}I-QNB, also add to the complexity. While ^{99m}Tc is the preferred SPECT tracer it does not have the luxury of simple substitution and the development of labelling methods for this nuclide represents

Table 1.2. SPECT tracer compounds and their applications.

Tracer	Application	Commercially available?
[^{99m}Tc]HMPAO	Cerebral blood flow	√
[^{99m}Tc]ECD	Cerebral blood flow	√
[^{123}I]IMP	Cerebral blood flow	√
[^{123}I]IBZM	Dopamine D_2 receptors	√
[^{123}I]β-CIT	Dopamine transporter	√
[^{123}I]Iomazenil	Benzodiazepine receptors	√
[^{123}I]QNB	Muscarinic cholinergic receptors	
[^{123}I]Dexetimide	Muscarinic cholinergic receptors	
[^{123}I]MK801	NMDA receptors	
[^{123}I]α-methyl-tyrosine	Amino acid transport	
[^{201}Tl]Chloride	Brain tumours	√
[^{111}In]Pentetreotide	Somatostatin receptors in tumours	√
^{133}Xe	Cerebral blood flow	√
[^{18}F]FDG	Glucose metabolism (using special SPECT camera)	

one of the major goals of SPECT research, promising advances being made in some areas[5].

Some of the key tracers available for SPECT imaging of the brain are shown in Table 1.2. Many are available from commercial radiochemical companies, either as the final product or in the form of a simple kit which is easily reconstituted on site. Some, however, are only available through specialized hospital-based radiopharmacies. The clinical applications of these products are described in other chapters.

Radiopharmaceuticals for imaging cerebral blood flow

Although SPECT imaging systems have been with us for well over 20 years, it was only when the cerebral blood flow imaging agents [^{123}I]IMP and particularly [^{99m}Tc]HMPAO became available that the technique gained wide acceptance. Since the imaging of cerebral blood flow still accounts for more than half of the clinical and research work carried out in SPECT it is worthwhile examining the properties of these tracers in more detail.

Two different classes of tracer are used to image cerebral blood flow: those which are metabolically inert and those which undergo a modification within the brain. The early cerebral blood flow methods all employed highly diffusible inert gases, such as non-radioactive nitrous oxide and radioactive ^{133}Xe. When such a tracer is introduced as a bolus into the circulation, equilibrium is soon established between the concentration of the tracer in the blood and in the brain. The rate at which the tracer is subsequently washed out of the brain gives a measure of cerebral blood flow. Ideally, in order that the equilibrium is rapidly established, the tracer should be lipophilic and therefore be able to pass with no hindrance from blood into brain and vice versa. Although ^{133}Xe is capable of providing absolute measurements of cerebral blood flow and is still used in some specialist centres, its limitations of fast washout and low energy make it unsuitable for high quality SPECT imaging. This prompted the search for an ideal tracer for SPECT imaging with the following characteristics: (i) uncharged, lipophilic and of low molecular weight to allow passive diffusion across the blood brain barrier during first pass through the brain. (This is a difficult balance to achieve since higher levels of lipophilicity often lead to increased binding to plasma proteins); (ii) rapid trapping and retention within the brain with a fixed distribution to allow sufficient time for the acquisition of the SPECT images; (iii) labelling with ^{123}I or preferably ^{99m}Tc.

The first agent designed specifically for imaging was [^{123}I]IMP (^{123}I-iodo-amphetamine)[6]. This has a high brain extraction and its initial uptake is linearly related to cerebral blood flow over a wide range of flows. It is then retained within the brain due to its conversion to hydrophilic metabolites and an affinity for non-specific protein binding sites[7]. However, the total activity in the brain increases markedly over the first 20 min[8] due to clearance from the lung and recirculation to the brain. Redistribution also occurs over a longer time period, resulting in a change in grey/white matter contrast. Although IMP has quite attractive kinetics, the cost and availability of ^{123}I will always restrict its clinical use.

The quest for a lipophilic agent which could be labelled with ^{99m}Tc centred on propylene amine oxime (PnAO)[9] but, like ^{133}Xe, this was found to have a very rapid washout, requiring very fast imaging. Subsequently a series of PnAO analogues were evaluated and the breakthrough came with the development of demethylated derivatives[10]. In particular the RR,SS(*d,l*) diastereoisomer of HMPAO (CERETEC™, Amersham International) exhibits close to ideal characteristics with long-term retention[11]. The retention in brain has been linked to an enzymatic reaction with glutathione, which converts the complex to a less lipophilic form, thereby trapping it behind the blood–brain barrier[12]. Some authors dispute the role of glutathione and propose binding to a protein in the cytoplasm as the mechanism[13]. The kinetic profile of [^{99m}Tc]HMPAO is as shown in Figure 1.3. The activity in brain rises rapidly in the first 1–2 min after administration, experiences a slight fall and then retains almost a fixed distribution with washout occurring at a rate of less than 2%/h for many hours thereafter[14]. These features make [^{99m}Tc]HMPAO a superbly practical tracer.

Its distribution in the brain is determined during the period from about 15 s to 1 min during intravenous injection. The scan, which can be carried out at any time within the next few hours, therefore measures the cerebral blood flow pattern at the time of administration and not at the time of scanning. This enables [^{99m}Tc]HMPAO to image cerebral blood flow during transient conditions or psychological challenges (see Chapters 2 and 5) without the patient having to undergo the scanning procedure at the same time. In conditions such as epilepsy this confers an enormous advantage to SPECT over PET.

[^{99m}Tc]HMPAO decomposes in vitro after reconstitution and has to be used within 30 min of preparation. Techniques for stabilizing the complex are now available and extend the life of the product for several hours[15].

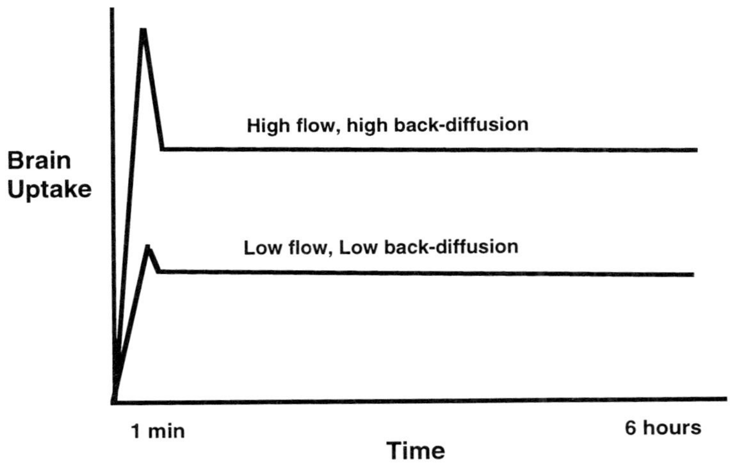

Figure 1.3. Brain uptake of HMPAO.

Although brain uptake is flow dependent, the first pass extraction also varies with flow, as does back-diffusion to blood. When cerebral blood flow is low there is sufficient time during transit of the ^{99m}Tc HMPAO through the brain capillaries for it to be extracted and trapped efficiently. As the flow increases, however, the transit time becomes less and a lower proportion of the tracer is extracted. In addition, some of the tracer diffuses back into blood before trapping occurs. The relationship between blood flow and [^{99m}Tc]HMPAO brain uptake is, therefore, non-linear, with the uptake less than expected at higher flows. This is of little consequence in routine imaging but where quantitation is required Lassen et al.[16] have proposed a correction algorithm for estimating CBF relative to a reference region:

$$\mathrm{CBF\%} = 100[(\mathrm{C/Cr}) \times a]/[1 + a - (\mathrm{C/Cr})]$$

where C is the count rate in the region being studied, Cr is the count rate in the reference region e.g. cerebellum, and a is a constant between 1.5 and 2.0. Inugami et al.[17] found the highest correlation between HMPAO and the $^{15}CO_2$ steady state PET technique using a value of about 2.0 for *a*.

The validity of [^{99m}Tc]HMPAO as a cerebral blood flow tracer has been thoroughly tested and compared with other techniques in normal and pathological conditions. Autoradiographic studies have compared HMPAO with iodo-antipyrine (IAP) in normal[18] and ischaemic[19] brain and clinical comparisons have been made with ^{133}Xe washout[20] and PET[21].

Radiopharmaceutical regulation

Radiopharmaceuticals are among the most highly regulated materials administered to patients because they are controlled both as medicinal products and as radioactive substances. All commercially available tracers undergo approval procedures to obtain a product licence for the purpose for which they are intended, a process which is both lengthy and time consuming[22]. Where complete 'in-house' synthesis is performed the radiopharmacy must hold a special manufacturing licence. Radiopharmacy facilities must also comply with a twin set of standards covering pharmaceutical and radiation safety. The aims in designing a radiopharmacy are

- to protect the product from external contamination;
- to protect the operator from radioactive and biological contamination;
- to protect the environment from radioactive and biological contamination.

All work must be carried out in accordance with good pharmaceutical practice, usually in high specification laminar flow cabinets. Quality assurance is vitally important[23], although the use of radioactive materials can give rise to unique problems. For example, while sterility is a prerequisite for intravenous tracers, preparations using radioisotopes such as ^{99m}Tc, which have half lives of only a few hours, have to be used before sterility testing can be carried out. Testing in this case is retrospective and is designed to give information about the preparatory technique in general. Radiation protection is often the most daunting feature of a large radiopharmacy, which at any time may hold 800 GBq of ^{99m}Tc generators, 100 GBq of other imaging nuclides and large quantities (e.g. 25 GBq of ^{131}I) of radionuclides for therapeutic use.

Interaction of radiation with matter

Although the information conveyed by a SPECT scan is intrinsically related to the pharmaceutical employed it is the interaction of radiation with matter that determines the fundamental sensitivity and limitations of SPECT.

All of the radiations and particles emitted during radioactive decay are capable of dislodging electrons from the electron shells of atoms, thereby causing ionization. The processes leading to and resulting from ionization have both beneficial and detrimental consequences. First, ionization causes disruption in some molecules and, consequently, has potentially harmful biological effects. The absorbed radiation dose is discussed below with reference to all of the types of radiation emitted. Second, as photons pass out of the body they can interact with the tissues themselves and the relationship between the measurement of the detected radiation and the amount and location of the radiopharmaceutical is altered. Finally, every type of radiation detector, including the simple film badge, uses the process of ionization to great advantage to detect the radiation.

Measurement of the detected radiation

Photons emitted by SPECT radionuclides interact with the electrons in the atoms of body tissues in two different ways (Figure 1.4). They can pass all of their energy

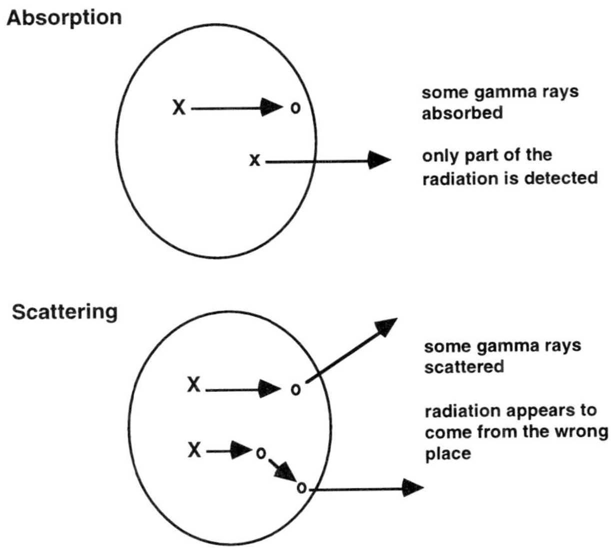

Figure 1.4. Absorption and scattering of emitted radiation.

on to the electron and thereby be completely absorbed in the tissue. Alternatively they can pass only part of their energy to the electron and in so doing be deflected to a different trajectory. These two processes are known as photoelectric absorption and Compton scattering.

Absorption results in a reduction of the detected signal, particularly for radiation originating in deeper anatomical structures. Most (but not all) SPECT systems therefore incorporate a factor known as the absorption or attenuation correction in the image reconstruction algorithm (see Figure 1.7). The correction usually assumes uniform attenuation throughout the brain and skull, although some novel SPECT systems obtain more accurate individual estimates of attenuation by simultaneous measurement of the transmission of gamma or X-rays through the head. The calculation and use of attenuation corrections is still a subject of much research and debate[3,24].

Scattering changes the trajectory of the photon, and consequently the spatial information in the signal will be misleading. The energy of the photon is also changed in the process, being reduced by an amount which is dependent on the angle by which the trajectory has been altered. The change in energy is used to compensate for scatter by using an energy analyser which rejects all photons whose energy is outside well defined limits known as the energy window. For example, when detecting 140 eV ^{99m}Tc gamma rays a common window setting is 125–155 keV, which excludes most photons which have undergone a major change in trajectory. Despite this 30–40% of all counts in a SPECT image are from scattered photons. Use of an energy window has the additional benefit of reducing the amount of background radiation included in the image.

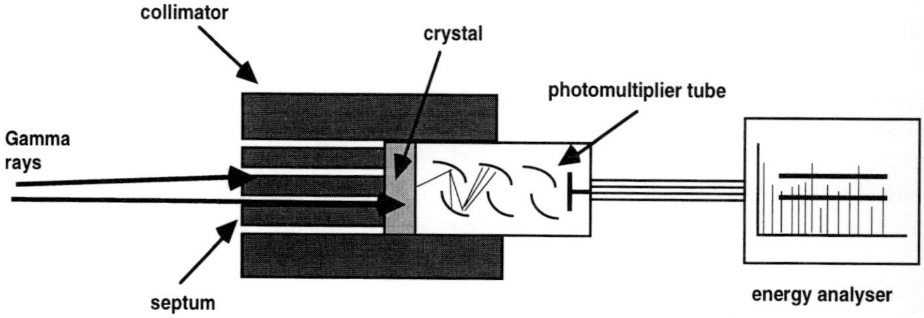

Figure 1.5. Scintillation detector.

Considerable efforts continue to be directed at methods of correcting effectively for the scattered component[24–26], including the use of dual energy windows, where data from the 'scatter' window is subtracted from that of the main energy window. In some cases scattering, which also varies with depth within tissue, is corrected for within the image reconstruction process at the same time as attenuation.

Radiation detectors

All SPECT systems use one of the simplest and most efficient types of radiation detector known as a scintillation detector. This has been in use since the early 1900s, when the scintillations, or tiny flashes of light, which are produced in certain materials when struck by radiation were counted by an investigator sitting in a darkened room. Nowadays modern electronics and computers have replaced this laborious task but the principle remains unchanged.

Briefly, the detector (Figure 1.5) consists of a crystal of sodium iodide, a photomultiplier tube and an electronic energy analyser and counter which cuts out the signals from photons with the wrong energy, e.g. scattered photons. Scintillations in the crystal result from ionization when the photons are absorbed, via the photoelectric and Compton effects, and as the ensuing electron orbital vacancies are filled by other loosely bound electrons a multitude of practically simultaneous tiny flashes of light are produced. Pure crystals of sodium iodide only scintillate at very low temperatures but when small amounts of impurities, known as activators, are introduced into the crystal scintillation occurs at room temperature. The activator used with sodium iodide is thallium and the crystal is referred to as NaI(Tl). The scintillation light is converted into a beam of electrons (and thus an electric current) by a photomultiplier tube. Large crystal assemblies, such as are found in gamma cameras, contain many tens of photomultiplier tubes which provide separate signals from different parts of the crystal.

The final component in a detector assembly is the lead collimator which surrounds the crystal. On its own a scintillation detector supplies no information on the origin of the radiation being detected, since photons striking the crystal from any direction produce a signal. Attaching a lead collimator as shown in the diagram above prevents photons other than those arriving parallel to the holes in the collimator, from reaching the crystal. Even when the angle to the holes is quite small the protons will still strike the lead walls (septa) rather than pass through to the crystal. The collimator therefore gives the detected signal directional information. In conventional planar gamma cameras the collimator directionality and the number and position of individual photomultiplier tubes combine to form an image with two-dimensional spatial localization.

Different collimator designs are available for different applications, with variations in the number of holes, their length and orientation (e.g. parallel, focusing and fan beam) and their septal thickness.

Tomographic imaging

Imaging with single photon emitting radionuclides presents particular difficulties compared with X-ray tomography or even positron emission tomography. The usual SPECT image, for example, is based on the detection of around 10^6 photons, whereas an X-ray CT may have as many as 10^{11} photons available for the same axial slice. In X-ray CT the position of the X-ray source and the position of the detector are known from the design of the scanner, thereby providing well-defined spatial information and allowing the reconstruction process to concentrate on determining the photon attenuation coefficients on which the final image is based. In contrast, with SPECT, not only are the attenuation coefficients unknown but so too are the positions of the sources of radiation, which must be inferred from the positions at which they are detected. As described above lead collimators are employed for this purpose, but these have considerable disadvantages, not least the fact that they absorb most of the photons reaching the detector. The collimator also provides less than perfect directional information and adds significantly to the overall resolution of the system. It is of interest to note that a collimator is not required on a PET scanner because coincident detection of the two gamma rays defines a line along which the positron annihilation must have occurred, therefore giving spatial information conceptually similar to X-ray CT. The absence of a collimator means that the sensitivity of PET is one to two orders of magnitude greater than that of SPECT[27].

SPECT is a technique, therefore, which has several disadvantages over other imaging modalities, notably an intrinsically lower sensitivity and consequently a lower signal to noise ratio. The reconstruction algorithms are also more complex, and to provide true quantitation the problems of correcting for tissue attenuation and photon scattering must be addressed. Despite these apparent limitations the overall performance of many of the SPECT scanners described below compares favourably with many PET scanners[27].

The fundamental requirement for the formation of a tomographic image is that data are collected from a number of different angles around the body. Using a single gamma camera head as an example, a series of conventional planar images

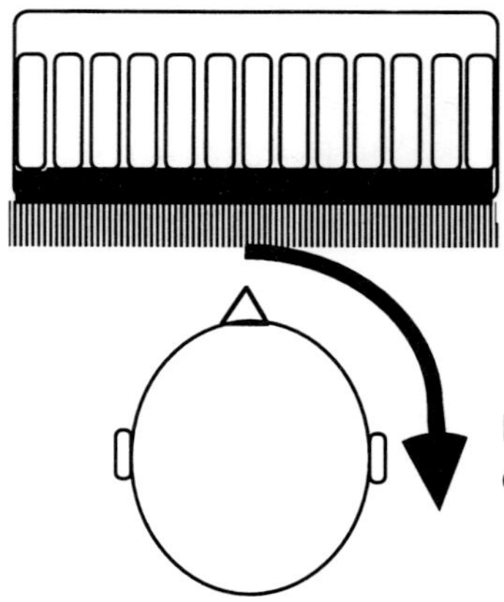

Figure 1.6. Operation of a single head gamma camera.

is taken, at equal angular spacing, through a 360° rotation (Figure 1.6). The rotation can take the form of a number of small discrete steps or maybe one continuous motion. The time spent at each position is very short, and each image is of poor quality. When all of them are combined, however, sections of the brain in axial, sagittal or coronal orientation are obtained. The orbit of rotation of a gamma camera head is crucial to the quality of the final image and should at all times be as close as possible to the patient. This maximizes the collection of photons as well as optimizing the resolution. Most camera systems now perform elliptical orbits.

Image reconstruction

Once the full data set from multiple angles has been obtained all of the data are combined mathematically in an endeavour to reproduce the distribution of the radionuclide, a process known as image reconstruction. X-ray transmission tomography (CT scanning) and emission tomography (PET or SPECT) use similar mathematical methods for the reconstruction of cross-sectional images. The two methods which are commonly used are filtered back-projection and iterative reconstruction.

In simple-back projection (Figure 1.7) the acquired signals (for a rotating camera) are the two-dimensional planar images in which the photons from all structures along lines perpendicular to the camera face are superimposed. There is no depth information in these images, the radiation from structures at different depths being overlaid. Reconstruction is simply the reverse of the acquisition process. An estimate of the original distribution is obtained by taking each image and back-projecting it onto an image matrix at the angle at which it was originally acquired. Each point in any back-projected line is assumed to have the same value, equal to its corresponding activity on the planar image. High activity in the original distribution, indicated by the black circles in the diagram, is reproduced

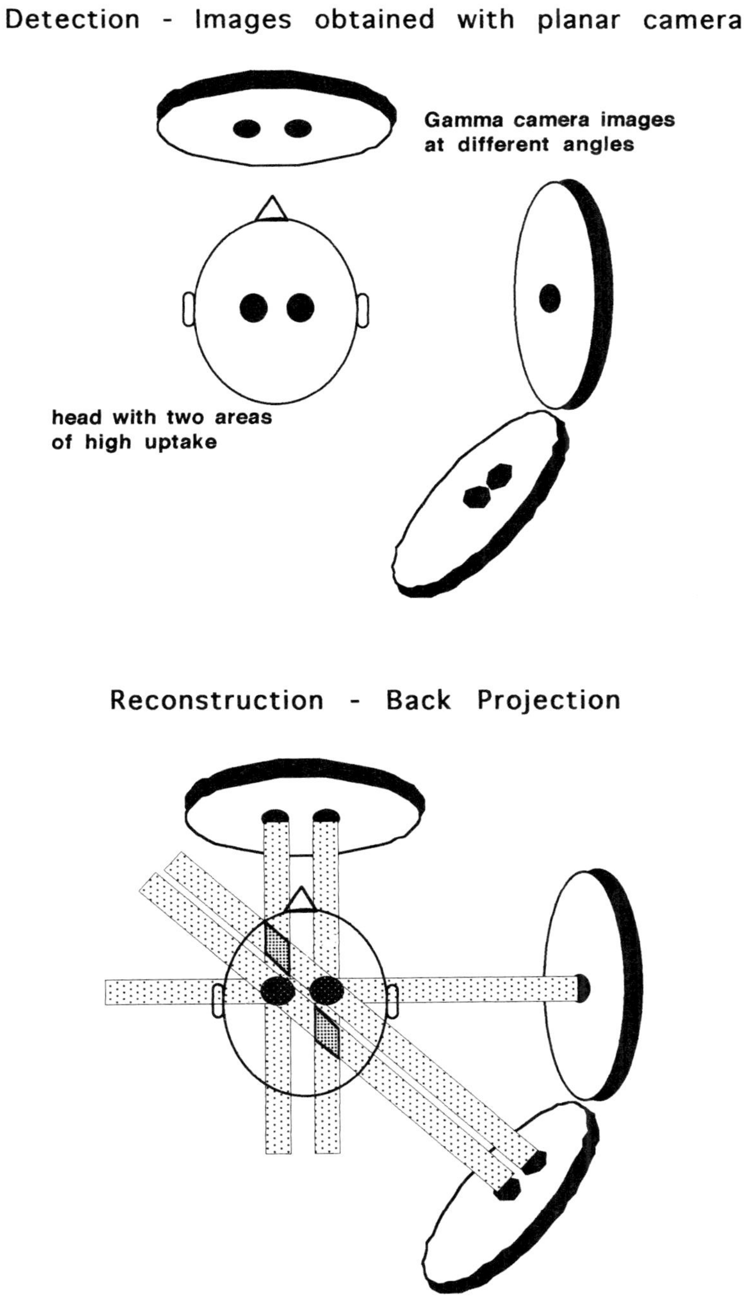

Figure 1.7. Reconstruction using back-projection.

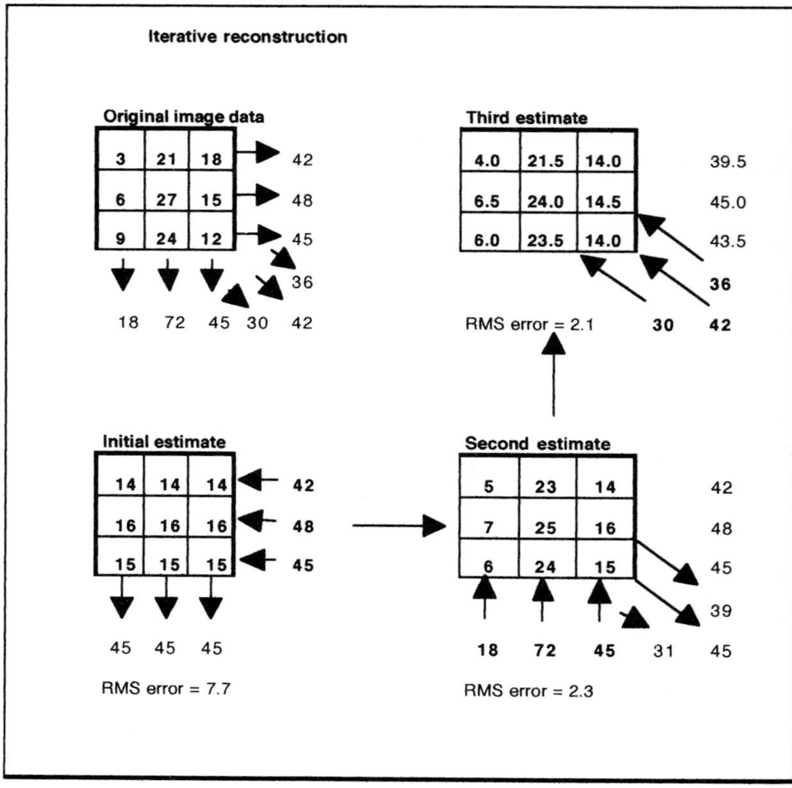

Figure 1.8. The iterative reconstruction algorithm.

when the back-projections from each angle overlap. Back-projection inevitably results in a high background activity in the reconstructed image, even extending outside the organ being studied, since the projections are smeared across the whole field of view. This causes blurring of the image. There is also a degree of structure to the background, resembling a star, due to the finite number of projections acquired. Both of these effects can be compensated for to some extent by filtering the data.

Figure 1.8 is a simplified illustration of the iterative reconstruction process. In this diagram the original distribution is represented by a matrix of numbers. The sum of the rows, columns and diagonals (only one shown) correspond to the acquired projections on the planar images. To reconstruct the object from the projections an initial estimate of the distribution is made and then adjusted in a stepwise fashion using one projection at a time. In this illustration the initial estimate is made by dividing the projection data from each row equally among the columns. The column data now projects to 45, 45, 45. It should be 18, 72, 45 so the column entries are each altered by one third of the difference. The second estimate is much closer to the object distribution but there are still differences.

The fitting process can be repeated using the diagonal projections as shown in the third estimate. Notice that this starts to affect the row projection values but the closeness to the true object distribution improves, as shown by a reduction in the root mean square (RMS) deviation between the estimate and the object data. This iterative loop can be repeated several times and the RMS error will decrease even more but it will never reach zero.

Filtering

Whichever method of reconstruction is used, the data need to undergo a process of filtering to reduce statistical noise, to enhance structures and to assist the reconstruction process. The specifications of the particular filter used affects both the nature and quality of the final image.

Mathematically, filtering is most easily accomplished by modifying the spatial frequencies found in an image. The concept of spatial frequencies is fairly straightforward. The high frequency components define the edges, that is areas in an image where the signal changes rapidly. Lower frequencies occur where the signal changes more gradually. Filtering is achieved by multiplying each frequency by a certain factor; if a particular frequency needs to be totally suppressed the factor assigned to that frequency is zero.

As shown in Figure 1.7, back-projection introduces a high background into the image, thus effectively suppressing the high frequency components. Filtering is the reverse process in which higher frequencies are selectively enhanced – this type of filter is known as a ramp filter. Since SPECT has a relatively poor resolution, if the signal changes very rapidly from one data point to the next the only explanation is artefact or noise. The role of the filter in this situation is to progressively dampen signals with higher frequencies. Above a certain, cut-off frequency the signals are totally suppressed. A high cut-off frequency will result in very little rejection and an image which has high resolution, but is 'noisy' and may be difficult to interpret. A low cut-off frequency will result in a smooth image in which little anatomical structure is apparent (Figure 1.9; see also colour section). The choice of filter depends on the characteristics of the scanner, the object to be imaged and the number of counts available for image reconstruction. Many different filters are in use[28,29] in different SPECT systems. There is as yet no 'correct' filter, personal preference and experience still determining the choice in most cases. In the near future newer filtering techniques may bring more scientific rigour to this aspect of SPECT[30].

SPECT technology

The earliest description of a single photon emission tomography system was given by Kuhl and Edwards in 1963[31] and was based on an ingenious analogue recording device. The first computer based system was described in 1971[32], predating the announcement of X-ray CT. Unlike the acclaim which greeted the introduction of X-ray CT and MRI, the enthusiasm surrounding the development

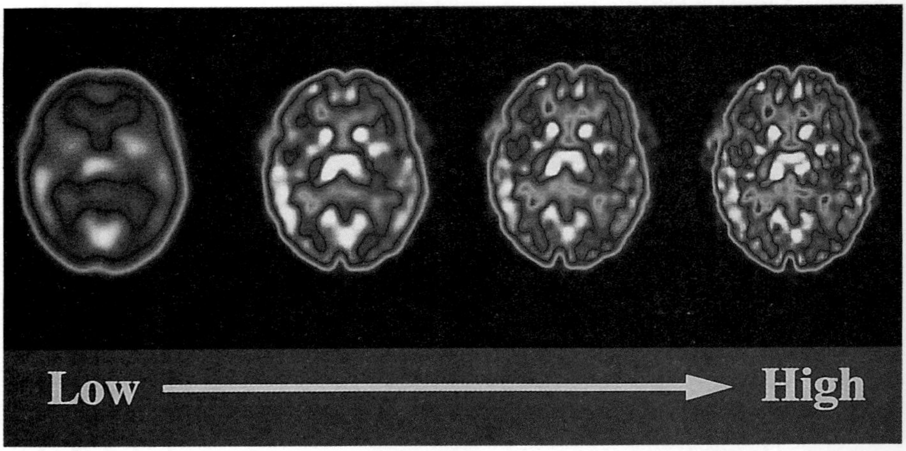

Figure 1.9. How the same raw data (in this case that making up the single illustrated orbitomeatal slice) may be filtered to different degrees to give different levels of resolution and 'noise' in the final image. (For colour version of this figure see page 179.)

of SPECT has been more muted, largely because of a lack of clinical applications in its early development. Understandably, the early years were marked by low quality images and rather troublesome machine artefacts, and even by the early 1980s little improvement had been achieved, prompting one supporter to remark in frustration that 'given poor hardware, sloppy technique and inattention to detail, the images will fail even to live up to the expectations of conventional scintigraphy'[33]. It is unlikely that this sort of criticism was on its own sufficient to stimulate the rapid developments in SPECT seen over the last decade: the development of new radiopharmaceuticals has been the life-blood of SPECT. Manufacturers have also responded positively and produced a range of novel SPECT system designs. Much praise is due to those manufacturers who ensured that the technology was readily available by marketing a conventional gamma camera system with the camera head mounted on a ring gantry. This accrued no significant extra cost and ensured that SPECT technology could be acquired by nuclear medicine departments without having to justify the expenditure on the basis of SPECT alone. This has provided a wide base of experience from which the present rise in interest stems.

These technological advances can be summarized by the improvement in resolution of SPECT systems over the last two decades, falling from about 20 mm to almost 7 mm for state of the art systems today.

Types of SPECT imagers

Some of the different types of SPECT imagers available are summarized below (Table 1.3)[34]. They encompass a large range of design philosophies to maximize the number of photons detected by a combination of reducing the distance between

Table 1.3. Types of SPECT imagers.

Manufacturer	Model	Detectors	Comment
Single-head gamma cameras			
Many manufacturers		Single head	Good for planar imaging Low quality for SPECT Prolonged imaging time
Multi-head gamma cameras		*Camera heads*	
ADAC	Vertex Plus EPIC	2	
	Solus EPIC	2	
Elscint	Helix	2	Multiple heads provide
	Varicam	2	increased sensitivity and
General Electric	Maxxus	2	better SPECT resolution.
	Neurocam	3	
Hitachi	V250DSP	2	2 headed scanners are more
Park Medical Systems	ISOCAM II	2	practical for general nuclear
Picker	PRISM 2000	2	medicine and whole body
	PRISM 3000	3	imaging
Siemens	MultiSPECT 2	2	
	MultiSPECT 3	3	
Sopha	Sophy DST-XL	2	Some 2 headed scanners
Toshiba	GCA 7200	2	also allow variable camera
	GCA 9300	3	angle geometry for
Trionix	BIAD XLT	2	180° cardiac SPECT
	TRIAD XLT	3	
Dedicated multi-detector SPECT			
Digital Scintigraphics	CERASPECT	Annular single crystal 63 PMTs	Brain only
Hitachi	SPECT 2000H	4 camera heads	Brain only
Medimatic	Tomomatic 64	4 banks 64 detectors	Brain only 5 slices
	Tomomatic 864	variable aperture	All organs
Shimadzu	Headtome	3 rings 128 detectors/ring	Brain only 3 slices
	SET 030W	3 rings 128 detectors/ring	All organs
Strichman Medical	SME810	12 focused	Brain only Single slice

the brain and the detectors and increasing the total detector area. Many of the most recent systems are capable of imaging positron-emitting radionuclides.

Conventional single head gamma camera

This design (Figure 1.6) was first demonstrated by both Jaszczak[35] and Keyes[36] and the great majority of SPECT systems presently in use are of this particular type. The gamma camera head is mounted on a rotating gantry and at each incremental step in the rotation a conventional planar image is acquired. While images of the heart can be obtained using a limited arc, a complete 360° rotation is required for the brain. The closer the camera head can be brought to the patient the better the image quality and many innovations have been tried, including the use of elliptical orbits and specially shaped camera heads to avoid the shoulders. Typically a scan involves 64 acquisitions taking a total of 40–50 min, throughout which time the patient must remain as still as possible. While the possibility of the patient moving during the relatively long scanning period and thereby corrupting the acquired data is a major disadvantage, one attraction of gamma camera systems is that truly three-dimensional data are produced, allowing sections in any orientation to be displayed with essentially the same resolution.

The artefacts which dogged rotating gamma camera systems in the early stages, such as field uniformity and mechanical instability, have now been largely resolved by the manufacturers or can be overcome by careful quality control procedures. The fundamental trade-off between sensitivity and resolution still exists, however although recent papers have demonstrated that the attitude to the use of high resolution collimators has been unnecessarily cautious and that the loss of sensitivity in using these is usually acceptable[37,38]. Fan-beam and cone-beam collimators enhance performance further, especially in the brain.

Although single head gamma cameras cannot match the performance of the systems discussed in the next two sections they have contributed much to the recent development of SPECT and must not be underestimated in routine static imaging, particularly of cerebral blood flow. One important limitation is the inability to image tracers whose distribution changes even relatively little in a time scale of the order of 40 min or less[39], which is the case with many of the newer radioligands.

Multiple head gamma cameras

In simplistic terms the use of multiple gamma camera heads mounted on a suitable rotating gantry increases the sensitivity of the system in proportion to the number of heads. In practice the true gain in sensitivity may be higher than this if the newer designs of converging fan and cone beam collimators are also incorporated, although it could also be less if the design of the heads prevents them being positioned as close as possible to the patient's head (this applies particularly to some triple-headed systems). There is no doubt that the quality of images from these new systems is markedly superior to those obtained using single handed scanners. The increased sensitivity of multiple head systems can be applied in two different ways; either by improving resolution or by reducing acquisition time, making imaging of dynamic processes possible and patient movement less important. Recently, novel work on improving the quantitative accuracy of SPECT as well as its image quality has taken advantage of the multiple heads to measure photon attenuation coefficients, which can then be used to correct image data. For example, Jaszcak[40] has shown that one of the heads in a triple headed system can be used to acquire transmission CT data from an external ^{99m}Tc source while the other two heads acquire emission data from the administered radiopharmaceutical, allowing the use of measured rather than estimated attenuation corrections.

Dedicated multi-detector SPECT systems

A number of systems have been designed specifically for brain SPECT and most of these acquire data in the same manner as X-ray CT i.e. by scanning a single or a few axial sections, moving the patient couch by a small distance, and repeating the process until the whole brain has been covered. The whole set of axial slices is then integrated to form a 3D data set and re-sliced in orthogonal orientations. One of the earliest machines using such a design was the Harvard scanner, which uses unique focused collimators. It also has an unusual detector motion which is tangential to the field of view, rather than a circular rotation.

The in-plane resolution and sensitivity of these systems are very good and probably represent the state of the art at the present time. The resolution in the axial direction depends on the slice thickness, which itself depends on collimator design, and slice spacing during the acquisition. By careful choice of parameters the resolution will be the same in each of the three dimensions. A major advantage of limited slice systems is that the acquisition time for a single or a few sections can be as little as 1–2 min, allowing useful sections to be obtained from difficult patients, such as children, provided patient movement can be restricted for this length of time. With patience, and accepting some movement, a set of sections can be built up in much the same way that CT scanners are used in practice. Of course, where movement does occur, true 3D reformatting is not possible.

The speed of multidetector systems confers another advantage, the ability to perform dynamic imaging. One particularly rigorous test of this capability is the quantitative imaging of cerebral blood flow with ^{133}Xe. Few systems are able to achieve this. The Tomomatic systems have been designed specifically for the purpose, and have four banks of detectors at right angles to each other, containing a total of 64 detectors. Each detector has up to five collimators, defining up to five slices in the head. Movement of the couch is again used when required to fill in the gaps left by the five slices. Another system which is capable of ^{133}Xe measurements is the Ceraspect, which has a single large, annular crystal which is stationary during operation. To provide tomographic data a series of three annular collimator sets rotate concentrically to the crystal. A total of 63 photomultipliers, arranged in three rings, view the crystal.

Choice of SPECT system

It would be ideal if all SPECT systems were capable of high resolution, high quality, dynamic imaging, but clearly a compromise suited to the particular user's own needs inevitably must be reached. This compromise often means that the most advanced system is rejected in favour of a more practical design, and rightly so. Purchasing a SPECT system involves a well considered balance between the basic performance, the expected workload, the organs to be imaged, the ability to undertake fast imaging and the user friendliness of the computer workstation. The recent exciting developments in radiopharmaceuticals should, however, be enough to entice potential users to commit the slightly higher resources needed to acquire high quality systems.

To maintain consistent image quality certain quality control and calibration procedures are required for SPECT systems. These vary from system to system, and may in some cases be time consuming. They will not be discussed in detail here, beyond stressing their importance.

The radiation dose delivered by radiopharmaceuticals

An important concern in any investigation involving the administration of radioactive materials is the radiation dose given to the patient. Many of the limitations of SPECT in terms of its resolution and its relatively long imaging times could be overcome simply by increasing the amount of radioactivity administered, but this would increase the dose of radiation delivered to the patient's body. Conversely any attempt to restrict the dose inevitably results in a degradation of image quality. The administered activity must represent a careful balance between the quality of the image and the radiation dose. Of course, the potential hazards arising from any use of radiation must be justified by the benefits accruing to the patient from the diagnostic information obtained from the investigation. In order to make such judgements it is essential that standard methods of assessing radiation dose are available, and are considered for each type of investigation.

Radiations contributing to the radiation dose

The radionuclides used in diagnostic imaging emit one or more of the following during decay: gamma-rays, X-rays, electrons, beta particles and positrons. (No alpha particle emitters are used in diagnostic imaging.) Each of these radiations is capable of causing ionization in the tissues through which they pass and they all contribute to the total absorbed radiation dose. The radiations can be considered as either penetrating or non-penetrating: penetrating γ-rays and X-rays deposit only a fraction of their energy locally. Part of their energy is also deposited in adjacent, and even remote, tissues. Some of the energy is not absorbed at all by the body, and exits from it. It can then be detected and used to produce the SPECT image. Most beta-particles and electrons (and very low energy photons) are non-penetrating, and are absorbed within 1 mm of entering tissue. The radiation dose from these radiations is therefore localized to the region of uptake of the radionuclide.

Conventional methods of radiation dosimetry assume a homogeneous uptake of radioactivity in single organs or tissues and calculate a homogeneous absorbed dose based on this. This implies that the dose to every cell is the same as that average value. Recently, however, more interest has been shown in techniques of microdosimetry, which take account of the abundance of low energy electrons (most commonly Auger electrons, which originate from rearrangement of the electron shells of atoms rather than directly from nuclear processes) released by radionuclides. These particles have a very short range of up to a few micrometres, but usually less than 1 μm. Radiopharmaceuticals which concentrate around, or are targeted at, specific cell types may deliver to those cells a radiation dose markedly greater than that calculated from a homogeneous distribution[41–43]. In such instances the cellular dose, instead of being primarily due to gamma rays, becomes dominated by the dose from low energy electrons absorbed close to the site of uptake. Estimates of dose based on techniques such as microautoradiography indicate that cellular doses range from a few times to several thousand times the average macroscopic dose. In some instances this amounts to a dose of tens of Grays. Whether this concentration of the absorbed dose at specific sites has any implications regarding radiation safety is still open to question[42,44].

Calculating radiation dose

The calculation of radiation dose in radionuclide imaging is more complex than in X-radiology. Unlike X-rays, when the exposure of the patient is completed within a fraction of a second, a radioactive material passes through the blood stream to different organs and undergoes active or passive transport and trapping. It is released or broken down, over a period of time which may be hours or even days and is then excreted. The radiation dose therefore is delivered to the patient given over a protracted period, albeit at a much lower dose rate. In addition, an X-ray usually irradiates only the organs of interest and, to a much lesser extent, those immediately adjacent to them. Radionuclides, on the other hand, invariably deliver doses to the whole body, the highest dose often being given to a remote organ rather than the organ of interest. In summary, physical, kinetic and anatomical factors all affect the calculation of radiation dose delivered by radiopharmaceuticals.

Quantities and units used for radiation dose

The quantity of energy deposited in tissue per unit of mass is referred to as the absorbed dose. The unit is the Gray (Gy), defined as 1 J/kg. The old unit, which is still used in some countries, is the rad, which is 0.01 J/kg. Thus 1 Gy = 100 rad and 1 mGy = 0.1 rad. For medical imaging the Gy is a very large unit and doses to individual organs are normally expressed in mGy.

Some types and energies of radiation are more effective than others at inducing tissue damage for the same total amount of energy. For this reason another term, the equivalent dose (H_T) has been introduced and this is defined as:

$$H_T = \sum_R w_R . D_{T,R}$$

where w_R is the radiation weighting factor for radiation R and $D_{T,R}$ is the absorbed dose for that tissue and radiation. The unit is the Sievert (Sv). All photons and electrons have a radiation weighting factor equal to 1 and therefore in diagnostic imaging the absorbed dose and equivalent dose are identical. The only exception to this is Auger electrons (emitted by almost all radionuclides) which may have a factor as high as 40.

Table 1.4. IRCP weighting factors for tissue radiation dose.

Tissue or organ	Weighting factor (w_T)
Gonads	0.20
Bone marrow (red)	0.12
Colon	0.12
Lung	0.12
Stomach	0.12
Bladder	0.05
Breast	0.05
Liver	0.05
Oesophagus	0.05
Thyroid	0.05
Skin	0.01
Bone surface	0.01
Remainder	0.05

A further term, which was introduced by the International Commission for Radiation Protection (ICRP) to facilitate the setting of dose limits in occupational exposures, is the effective dose[45]. This uses weighting factors for individual organs and tissues, with each weighting factor representing the fraction of the total stochastic radiation risk (the risk of fatal cancer and serious inherited disorders) attributed to that organ when the whole body is uniformly irradiated. The effective dose is then obtained by multiplying the dose equivalent by the weighting factor for that organ and summing over all the organs irradiated.

$$\text{Effective dose, E} = \sum_T w_T . H_T$$

where w_T is the weighting factor for tissue T. The unit of effective dose is again the Sv and most diagnostic investigations give doses of a few mSv. The tissue weighting factors published by the ICRP are given in Table 1.4. The 'remainder' constitutes the adrenal glands, brain, upper large intestine, small intestine, kidney, muscle, pancreas, spleen, thymus and uterus, as well as any other organ selectively irradiated.

The use of the effective dose is an attempt to provide a practical single value which integrates the doses from all organs when the body is non-uniformly irradiated and which then represents the same total risk as if the body had been uniformly irradiated at that dose level. Although it was never intended for use in radionuclide imaging it has been widely adopted as a convenient method of comparing the risk from different radio-diagnostic procedures by means of a single value. Although The Medical Internal Radiation Dose Committee in the USA has rejected the use of effective dose as inappropriate[46] there is sufficient justification to support a cautious adoption of the unit[47].

The administered activities and effective doses for some commonly used radiopharmaceuticals are shown in Table 1.5. The figures are intended to be indicative only: the limitations and judgements underlying their calculation make a more precise interpretation inappropriate for an individual patient. The effective doses

Table 1.5. Activities and effective doses of some commonly used radiopharmaceuticals.

Tracer	Effective dose (mSv/MBq)	Usual activity[a] (MBq)	Effective dose (mSv) for usual activity
[^{99m}Tc]HMPAO	0.0093	500	4.7[48]
[^{99m}Tc]ECD	0.0079	500	4.0[b]
[^{123}I]IMP	0.027	110	3.0[48]
[^{123}I]IBZM	0.032	120	3.8[b]
[^{123}I]Iomazenil	0.016	185	3.0[b]
[^{123}I]QNB	0.05	150	7.5[c]
[^{123}I]Dexetimide	0.0395	150	5.9[49]
[^{201}TI]Chloride	0.23	75	17[48]
[^{111}In]Pentetreotide	0.081	110	8.9[b]
^{133}Xe	0.001	400	0.4[48]
[^{18}F]FDG	0.02	300	6.0[48]

[a]See comments below on variations in activities administered in different countries
[b]Estimated from manufacturer's product leaflet using weighting factors shown above
[c]Unpublished estimate

Table 1.6. Effective radiation doses from background radiation and some non-SPECT diagnostic procedures (from Ref. 51).

	Effective dose (mSv)
Diagnostic procedure[51]	
X-ray CT head	1.8
X-ray CT chest	8.3
X-ray CT abdomen	7.2
X-ray chest	0.04
X-ray barium meal	4.6
X-ray barium enema	8.7
Background radiation	
UK: Mean annual dose[50]	2.2
UK: range of annual doses[50]	1–100
Aeroplane crews: annual cosmic ray dose[51]	2.5–6.0
Miners (non-coal): annual dose from radon daughters[52]	14.0

from other radiodiagnostic procedures and from background sources of radiation are shown in Table 1.6 for comparison with these values.

Radiation effects and risks

Ionizing radiation can cause both deterministic and stochastic effects in irradiated tissue. Deterministic effects, such as skin burns, hair loss, induction of cataracts and radiation sickness, result from the killing of cells. They are associated with dose thresholds of at least a few Gray and do not occur at the dose levels encountered in routine medical diagnosis. Stochastic effects, such as the induction of cancer and genetic changes, are a consequence of non-lethal mutation of cells. The probability of a stochastic effect occurring increases with the dose delivered to the tissue. The question of whether stochastic effects occur at the dose levels used in SPECT is one which even the ICRP admits still has no directly proven answer. They state "there is little direct evidence of harm at levels of annual dose at or below the limits recommended by the Commission" and in formulating their recommendations they accept that "a good deal of scientific judgement is required in predicting the probability of harm resulting from low doses"[45]. Indeed, the possibility of a beneficial effect of low dose radiation (a phenomenon termed hormesis) cannot be ruled out by present evidence[53,54]. The balance of both theoretical and practical evidence[55], however, attests to a harmful effect of ionizing radiation at all dose levels and the approach of ICRP is aimed at using estimates that are not likely to underestimate the consequences of exposures[45]. The present estimate of the risk of fatal cancer is 0.05 per Sievert for the whole population. Risk estimates for other end-points have also been published[45].

Radiation dose limits

For workers exposed to radiation, including those in health care, a set of dose limits has been recommended by the ICRP[45]. These limits have found general acceptance by national regulatory bodies and they provide a consistent basis for the control of occupational exposures. No such dose limit has been set on medical radiation exposures. Instead, the ICRP made an unequivocal statement in their 1990 Recommendations[45] which has universal support in medicine. "Medical exposures are usually intended to provide a direct benefit to the exposed individual. If the practice is justified and the protection optimized, the dose in the patient will be as low as is compatible with the medical purpose. Any further application of limits might be to the patient's detriment. The Commission therefore recommends that dose limits should not be applied to medical exposures." Nevertheless, the ICRP encourages professional and regulatory bodies to introduce dose constraints. This form of restraint is achieved most readily by the use of guidelines on the levels of activities to be administered which, with present day technology, should be sufficient to produce images of diagnostic quality in a reasonable time. These constraints can be exceeded in individual cases, but use of higher levels on a routine basis would require justification.

In the UK this philosophy is very formally and strictly implemented through the Administration of Radioactive Substances Advisory Committee (ARSAC), who advise a maximum usual activity of 500 MBq for SPECT using [^{99m}Tc]HMPAO. A review of the literature, however, suggests that a consensus has yet to be reached regarding both the guideline levels and the formality of their implementation throughout the rest of the world. The use of the higher activities shown in Table 1.7 would be difficult to justify on a routine basis with a modern high quality SPECT system.

Dose limits in research projects

The system of constraints also applies in research. Generally, ultimate approval of a research project involving the administration of radiopharmaceuticals to

Table 1.7. Variation in administered activities given for HMPAO SPECT in different countries.

	Typical literature values for SPECT with [^{99m}Tc]HMPAO (MBq)
UK	500
USA	700–1400
Europe	750–1000
Australia	700–900

Table 1.8. Risk/benefit comparison for radiation doses administered during research projects.

Level of risk	Risk category	Effective dose (mSv)	Level of societal benefit
Trivial	I	<0.1	Minor
Minor to intermediate	II		
	IIa	0.1–1	Intermediate to moderate
	IIb	1–10	
Moderate	III	>10	Substantial

normal or patient volunteers lies with a local or regional Ethics Committee, although a separate application to a national body may also be required (e.g. ARSAC in the UK). Like all submissions to Ethics Committees the risks associated with the investigation must be weighed against the potential benefits. These considerations are necessarily complex and much advice has been issued to guide Ethics Committees[56,57]. Although it is possible to provide risk estimates for radiation (as shown earlier) which are infinitely more detailed than those available for virtually all other medical procedures or new pharmaceuticals, ICRP[58] have followed the lead given by the World Health Organisation in 1977[59] and suggest that research involving radiation be assessed with reference to a set of risk/benefit categories (Table 1.8).

An effective dose below 0.1 mSv represents a risk which the majority of people would consider as trivial and the justification required for using such radiation levels needs to be no more than increasing knowledge. For doses > 10 mSv, however, the potential benefits usually should be related to the saving of life or the prevention or mitigation of serious disease. Most SPECT investigations fall into the higher of the intervening categories and ICRP would expect a benefit directly aimed at the cure or prevention of disease.

SPECT image interpretation

The value of any imaging technique and its place in clinical practice and research is determined ultimately by the quality and rigour of the methods used to interpret the images. This is particularly valid for functional imaging techniques such as SPECT, where the process of interpretation entails an appreciation of the processes underlying the formation of the image together with the appropriate method of analysis of the imaging data. The expertise required encompasses many disciplines

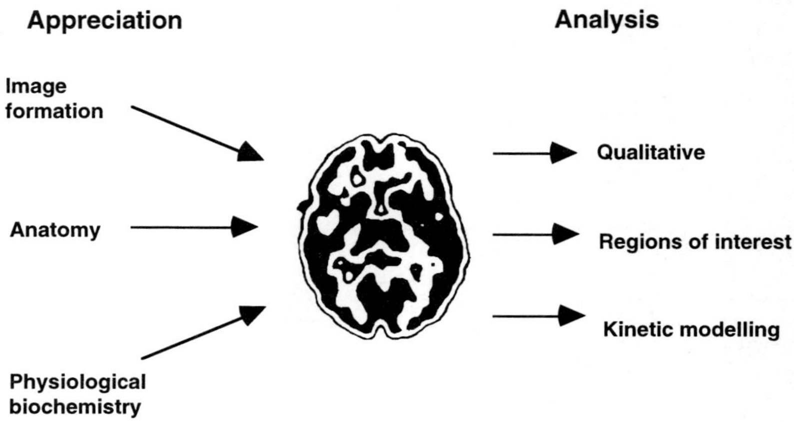

Figure 1.10. The elements of image interpretation.

(Figure 1.10) and the best results are often achieved where there is a multi-disciplinary input.

The influence of the physical processes involved in image formation

The appearance of SPECT images is clearly affected by all of the physical processes described above relating to the interaction and detection of radiation. The type of scanner, the detection method, the reconstruction and filtering algorithms and attenuation corrections all contribute to the individuality of images obtained from a particular system. Perhaps the greatest influence on the final images, however, is exerted by the resolution of the SPECT system. The precise neuro-anatomical localization which is possible with other radiological techniques clearly cannot be achieved with SPECT, which has a resolution around 1 cm compared to < 1 mm for MRI.

All imaging systems experience difficulties in imaging objects which are below their inherent spatial resolution (normally indicated by a term known as full width at half maximum or FWHM). The blurring caused by poor resolution affects only the edges of large uniform objects but as the object becomes smaller this blurring extends to the centre causing, in addition, an apparent drop in signal. Therefore a low signal detected on a SPECT scan can be caused by two factors, low function or thinning of anatomical structure.

The precise influence of resolution is dependent on the size, shape and relationship of anatomical structures as well as on the tracer uptake within each region. The brain, however, represents a complex model, being made up of many closely spaced and irregularly shaped structures and it is difficult to envisage how any particular SPECT scan will be affected. One possible way of investigating these effects is by computer simulation of the blurring[60]. Mazziotta used this technique with a pathological section of the brain but the use of MRI sections is a more

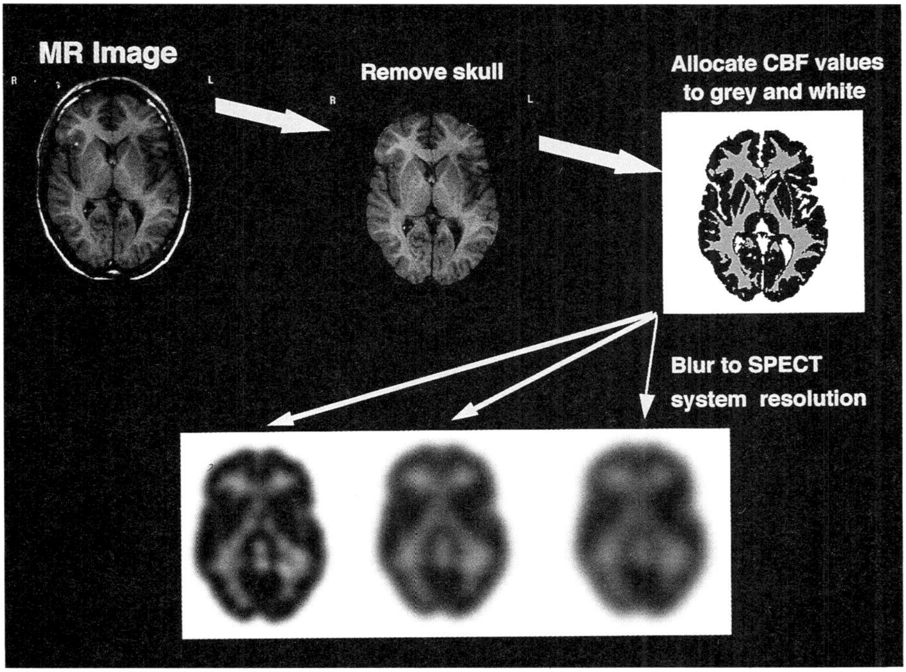

Figure 1.11. Simulation of a SPECT image of cerebral blood flow showing the effect of spatial resolution.

practical solution. Figure 1.11 shows the simulation of a SPECT image of cerebral blood flow for a tracer such as HMPAO and assumes uniform blood flow values in the grey matter, white matter and other tissues to be in the ratio of 2.5 : 1 : 0. Three different SPECT resolution values are used to illustrate the significant variation in visual appearance and demonstrate the potential individuality of any SPECT system. Simulations of other tracers, including receptor ligands, can be carried out in the same manner.

The influence of anatomy

SPECT images are not neuroanatomical maps, yet the knowledge of neuroanatomy required to interpret the images is at least as great as for other radiological techniques. In imaging cerebral blood flow, for example, it is not surprising that the images appear to demonstrate anatomical features since CBF is itself anatomically organized. In particular, there is a sharp contrast between grey and white matter due to the three-fold difference in CBF values between the two tissues. Cortical regions, subcortical structures (caudate, putamen, globus pallidus and thalamus), brain stem and cerebellum therefore are delineated quite readily from adjacent white matter on SPECT images of CBF.

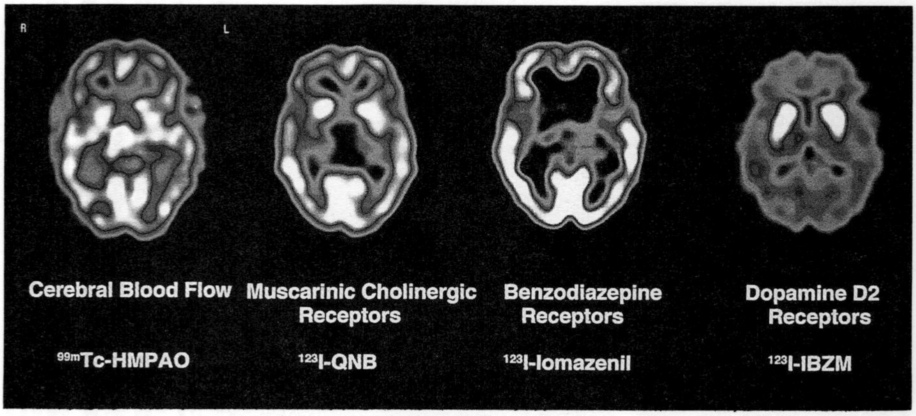

Figure 1.12. **SPECT images illustrating different uptake patterns obtained using different radionuclides. (For colour version of this figure see page 179.)**

Normal variations in neuroanatomy, especially changes with age, can have a great influence on SPECT images. The normal frontal to posterior gradient in cerebral metabolism (and CBF) disappears and may reverse with advancing age[61] since reductions in neuronal size and glial cell density are more marked in frontal cortex. Ventricular enlargement and sulcal widening are also part of the aging process, and have marked effects on SPECT images. In an ideal world structural imaging would always be performed in conjunction with functional imaging and would incorporate a correction for atrophy[62] or other structural anomaly or pathology (e.g. haematoma).

As the number of ligands available for imaging of neuroreceptor sites increases[63] so too does the need for an intimate knowledge of chemical neuroanatomy[64–66]. While an enormous amount of literature is available on neurotransmitter/receptor systems much of these data have been obtained from rat or primate brain, and uncertainty accompanies its extrapolation to humans. As far as localization is concerned, most neurotransmitters are distributed widely (but variably) in the cerebral cortex, except for dopamine which is confined to frontal and cingulate cortex. The greatest variation in distribution, however, is in subcortical structures and cerebellum. Figure 1.12 (see also colour section) shows four SPECT images produced using HMPAO to trace blood flow, and using ligands targeted at muscarinic cholinergic ([^{123}I]QNB), the benzodiazepine ([^{123}I]Iomazenil) and dopamine D2 ([^{123}I]IBZM) receptors. The absence of uptake in thalamus, the absence of uptake in all subcortical structures, and the predominance of uptake in basal ganglia, respectively, are consistent with the expected receptor distribution[67–69].

Expectations are high for the usefulness of neuroactive substances in SPECT imaging. These hopes will only be fulfilled if a clear understanding of both macroscopic neuroanatomy and chemical neuroanatomy underlies the interpretation of the images.

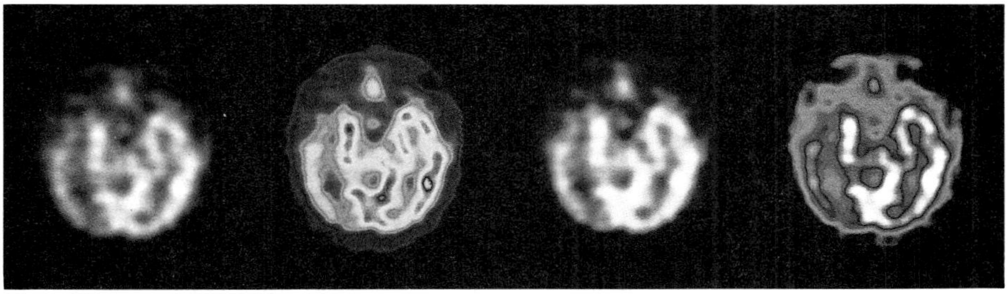

Figure 1.13. The same slice is displayed using four colour scales (left to right). In all cases, no signal is black. (a) Monochrome grey scale, with white showing the highest signal, black the lowest. (b) 'Rainbow' type colour scale using a combination of colours, rising through blue, green, yellow, red and white (only a very small area of white is seen in the posterolateral left temporal cortex. (c) Monochrome 'hot metal' type red scale, from dark red to light yellow/white. (d) Bichrome blue/red 'fire and water' colour scale, with reds, yellows and whites showing the higher signal levels. (For colour version of this figure see page 179.)

Qualitative interpretation

The results of clinical SPECT investigations are normally reported in the same way as CT and MRI, i.e. visually and qualitatively. This process relies on the type of expert knowledge discussed above and experience based on clinical feedback. Such practice is also now typical of many PET applications which have finally been released from the shackles imposed by inappropriate use of quantitation. It is not necessary, nor is it often effective, to resort to complex mathematical manipulations to derive relevant information from SPECT images. As Di Chiro and Brooks said "How sweet the siren song of mathematics that it blinds people to what is before their very eyes"[70].

Visual reporting, utilizing a solid foundation of neuroanatomical knowledge and manipulating the image data into different image planes, has the advantage that all the available data are considered without imposing the type of constraints associated with the size and positioning of regions of interest when quantitative techniques are used. Reporting should, however, be formalized to some degree, especially when the individuals interpreting images are relatively inexperienced. Blind interpretation by two or more investigators followed by a consensus conference is good practice, especially when the information is used to make important clinical decisions.

False colour scales are used universally in SPECT and enable a greater contrast range to be achieved with a single window setting. This is particularly important when images are transferred to a hard copy device for presentation. Many different scales have been used (Figure 1.13, see also colour section) and the choice is entirely one of individual preference, although the time taken to gain experience in the sensitivity of each scale should not be underestimated. Scales with rapid changes from one colour to another should be used with caution as they can enhance small differences which may not be significant.

In the succeeding chapters of this book you will see colour images presented

Table 1.9. Classification of region of interest methods of image analysis.

Analysis method	Description	References
Manual		
Geometric	Simple square or circular ROIs placed on specific structures or areas of abnormal or normal uptake	71
Freehand	Regions drawn by hand to include whole brain, hemispheres, specific structures or areas of abnormal or normal uptake	72, 73
Automatic		
Surface template	Automatically drawn ROIs which follow, for example, the external surface of the brain	74, 75
Threshold	Automatically drawn ROIs which encompass regions above or below a specific threshold value	76, 77
Directed		
Template method	Templates or irregularly shaped ROIs which outline consistent 'structures' underlying the functional image, for example, medial temporal or calcarine cortex	78
Anatomy or function guided	ROIs drawn manually on a specific structure or functional area identified on another scan or image modality and transferred to the co-registered functional image	79, 80
Pixel-by-Pixel		
Anatomy based	Functional measurements are expressed relative to a co-registered anatomical image	81
Function based	Functional measurements are expressed relative to an 'atlas' of functional data using techniques such as statistical mapping	82, 83

in the 'fire and water' blue/red colour scale, a 'hot metal' red scale, a single phase grey scale (all illustrated in Figure 1.13) and a biphasic grey scale.

Quantitative analysis of brain uptake

A variety of methods is available to enable quantitation of brain uptake. Each is based on recognizing a region (or volume) of interest made up of single or multiple image pixels. A classification of the most popular methods is given in Table 1.9.

An example of ROI placement using templates is shown in Figure 1.14, and an example of ROI placement according to co-registered MRI data is shown in Figure 1.15 (see also colour section). The choice of method should, of course, be suited to the application and although increasing emphasis is being placed on the use of pixel-by-pixel methods unnecessary complexity can be counterproductive. Many effective studies have used simple ROI measurements of absolute uptake, relative uptake or change in uptake.

Absolute measurements demand careful calibration of the SPECT system[24] and are less commonly used than relative uptake measurements. [^{99m}Tc]HMPAO

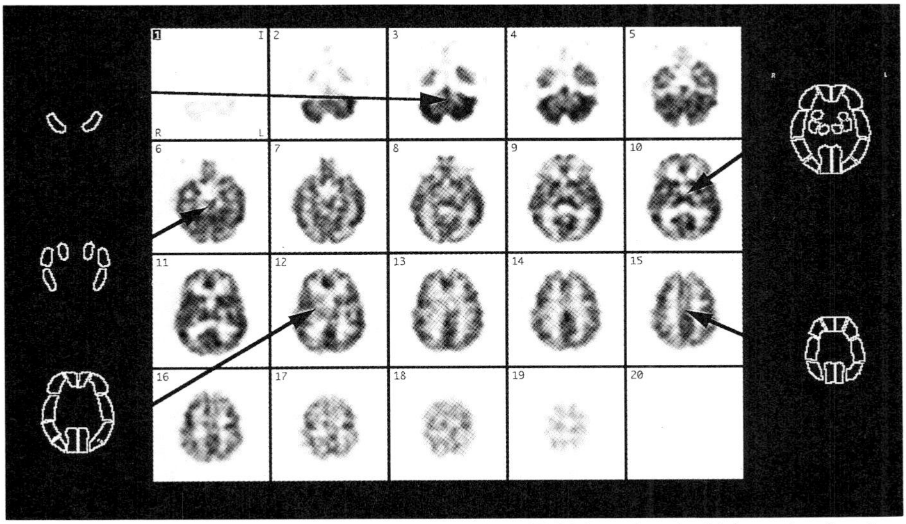

Figure 1.14. SPECT dataset displayed with examples of ROI 'templates' which can be placed on the appropriate slices. Most systems allow movement and some distortion of the ROI pattern, and some allow for movement of individual ROIs within the pattern. The more the pattern can be modified, the better it will fit different shapes of brain, but more subjectivity is then introduced into the analysis.

SPECT cannot readily produce absolute values of cerebral blood flow and therefore flow values are expressed relative to a standard reference region. The cerebellum is often used for this purpose, although if there is strong evidence that a certain region is unaffected or minimally affected by the disorder under investigation then that region can be used for normalization: the calcarine cortex, basal ganglia and frontal cortex have all been used. Where the choice of a single reference region is questionable the less sensitive, but more robust, method of normalizing to the average of a large number of regions (or the whole brain) can be adopted. In neuroreceptor imaging, receptor:ligand ratios are usually expressed as the uptake in the region with specific uptake of the ligand relative to that in a region with only non-specific uptake. For example, the cerebellum is used for muscarinic cholineric receptors[68] and the frontal or calcarine cortex for dopamine D2 receptors[84,85].

In studies involving functional activation, especially those using cerebral blood flow tracers such as [^{99m}Tc]HMPAO, quantitation of the change in uptake within particular regions can be achieved in a number of ways. One of the simplest is to express the change in terms of the z-score. This is given by

$$z = [C_r - C_m]/SD$$

where C_r is the signal from region r, C_m is the mean value for that region for all patients in the study or for a control group of patients and SD is the standard deviation associated with C_m.

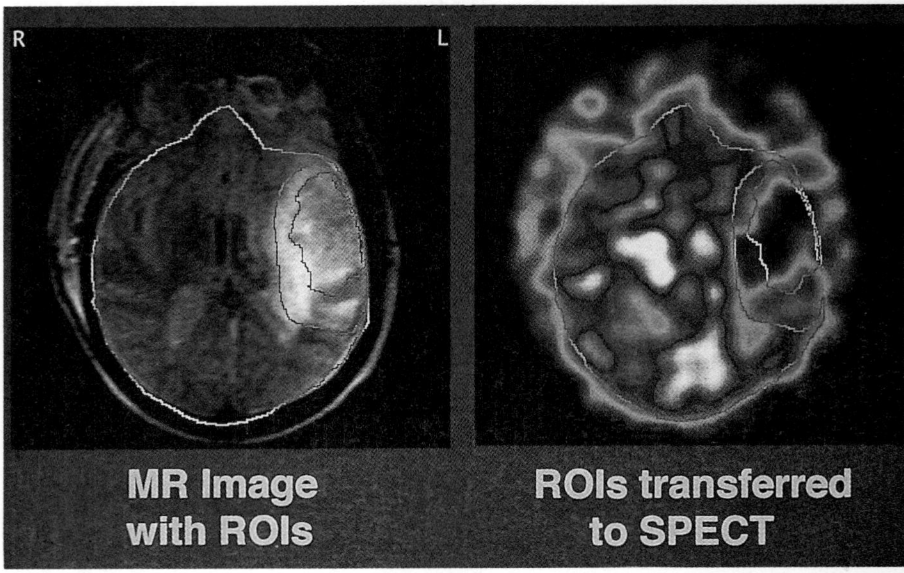

Figure 1.15. Placement of regions of interest using anatomical image data. The MR and SPECT images are co-registered. ROIs are drawn around anatomical features of interest on the MR images, and the corresponding SPECT slice is then redisplayed with the MR-determined ROI in place. (For colour version of this figure see pages 00–00.)

Although automatic methods of ROI generation avoid observer bias they can be unpredictable and a 'black-box' approach must be avoided. Even the simpler surface template and threshold methods must be carefully monitored, especially in abnormal scans when, for example, the underlying surface anatomy can be deformed or thresholding fails due to low uptake in the appropriate structure.

Regional analysis can utilize the apparent 'structural' information within the functional image, but the sophistication and rigour of the analysis is improved when true structural images are also available to provide a precise definition of underlying anatomy. Inevitably, anatomy-guided ROIs and pixel-by-pixel analysis methods require co-registration of image data sets. Numerous approaches to the co-registration of images within and across modalities have been used, a summary of which is shown in Table 1.10. Prospective techniques that define the position of the subject before the scan has been acquired and retrospective techniques that use information inherent in the images after the subject has been scanned have both been used to co-align images. Methods of standardizing images have been developed using anatomical atlases or plastic transformations and allow comparison between subjects rather than purely within a subject.

Patient immobilization methods attempt to fix the subject's head with respect to the scanner by means of a mask or some other rigid fixation device, but will always have limited ability to replicate subject positioning between different types of scanners. This can be overcome by the use of stereotactic frames which

Table 1.10. Classification of methods of co-registration of images.

Method	Description	Reference
Prospective	Patient immobilization	
	Fiducial markers (head or mask)	86
	Stereotactic frame (head or mask)	87
Retrospective	Point landmarks	78, 88
	Matching spatial moments of 3D volume	89
	Correlation of voxel density distributions	90, 91
	Matching of anatomical surfaces	92, 93
Standardization	Anatomical atlas	80, 94
	Plastic transformations	94, 95

are fixed to the patient's skull, as long as the frames are visible across modalities. To co-register the images, the co-ordinate transformation required to map one image onto the other can be calculated from the known dimensions of the frame. External fiducial marker co-registration is a less invasive method which depends on precise repeatability of positioning of the markers. The most popular methods are those involving voxel density distributions and anatomical surface matching techniques under the retrospective group heading. Landmark matching is an alternative to the use of external fiducial markers but may be time-consuming, especially in low resolution modalities such as PET or SPECT, because it requires the operator to identify and locate a sufficient number of points to obtain satisfactory registration of the images. Matching spatial moments of 3D objects is a method that calculates the affine transformation that will map one image onto another. It is really only suitable for within-subject and within-modality registration and requires an a priori definition of precisely corresponding subvolumes in each scan. Correlation of voxel intensity distribution uses an algorithm which calculates the ratio of one image to the other at each voxel in the image and then aligns the two images such that the variance of this ratio across all voxels is minimized. The technique is fully automated and is independent of the specific anatomical structures being imaged. Although this technique was originally developed for within-modality registration it has been modified so that PET onto MRI registration is also possible. 3D global shape properties of anatomical and functional image data sets can be used to calculate the required inter-scan co-ordinate transformation for image registration. Surface fit methods use a set of 2D contours of the brain surface to co-register images. One set of contours, usually from a high resolution modality (MRI, CT) constitutes one rigid body or head and is used as a reference image for another set of 2D contours that form another rigid body or hat. The 'hat' is then fitted onto the head using an algorithm that minimizes the volume between the two surfaces by an iterative non-linear least squares procedure, varying the translation, rotation and scaling parameters.

Standardization methods have been developed by several groups working in PET imaging, primarily to compare images across subjects performing the same activation task, the image data being summed together to produce better counting statistics for analysis. These methods usually involve aligning all scans to a similar orientation in order that they may be compared to a standard anatomical atlas. Variations in neuroanatomy are then accounted for by a 3D warping of the images to form a common average or to fit an idealized model of the brain. Algorithms for plastic deformation and warping of whole brains are both complex and somewhat controversial[96].

Functional neuroimaging is increasingly being driven by a combination of the standardization methods of image co-registration and the pixel-by-pixel based methods of analysis. In theory, by standardizing the images in these ways, regional quantification can be performed without bias or subjectivity for individual subjects or groups, either within or between modalities. The need for caution in adopting these techniques is highlighted by the continuing debate on their statistical basis[97] and by a recognition that the techniques are principally designed for functional activation studies and that their application in situations involving structural pathology remains limited[98].

Physiological biochemistry and kinetic modelling

Wagner[99] has recently used the term 'in-vivo topographical physiological chemistry' to describe emission tomography. The term not only conveys the potential utility of the technique but also its potential complexity. The amount of a radiolabelled tracer measured in the brain following intravenous administration is influenced by a multitude of physiological and biological processes. In SPECT the primary aim of imaging is to obtain information on one or more of these processes. To achieve this, it is often necessary to resort to a pharmacokinetic model which, by taking account of the presently understood properties of the tracer and assuming various idealized tissue compartments and kinetic parameters, is used to relate the (time varying) tracer uptake to the desired functional variable.

The boundless capacity for development of SPECT tracers is matched only by the number of possible pharmacokinetic models, and a full discussion of this subject is clearly beyond the scope of this chapter[100,101]. Clearly, some tracers such as HMPAO require only the simplest of models, as described earlier. It may be tempting to assume that simple empirical measurements, such as specific/non-specific count ratios, which are easily implemented in the clinical setting, are acceptable even for more complex tracers. Unfortunately, although this may be possible in some cases[102], this is generally a false hope[103] and if no account is taken of factors such as peripheral clearance, plasma protein binding, cerebral blood flow and regional non-specific binding, the results from this type of analysis will be at best misleading.

Some of the most complex, yet at the same time the most highly developed kinetic models apply to the field of neuroreceptor pharmacology[104–106]. A brief discussion of the principles involved is merited to illustrate the type of problem which PET scanning has addressed in enormous depth but which is only now being tackled in SPECT with the same rigour. Figure 1.16 shows a typical example of a kinetic model[107] for [^{123}I]iomazenil (a benzodiazepine receptor ligand) in the brain.

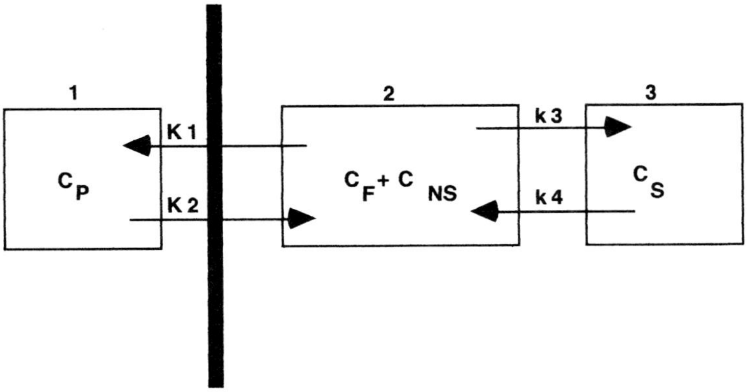

Figure 1.16. Kinetic model for the distribution of [^{123}I]iomazenil in the blood and brain.

This has been simplified to a three compartment model to describe the movement of the radioligand from blood into brain. The first compartment (C_P) is blood: only the concentration of unmetabolized ligand in plasma is considered. Within the brain the ligand exists in two separate compartments, the first of which contains both freely available (C_F) and non-specifically bound (C_{NS}) components, which are assumed to be in complete equilibrium. The second brain compartment (C_S) is the specific, receptor bound component. Rate constants describe the rate at which ligand can pass from one component to the other. K1, for example, describes the rate at which ligand enters brain from the vascular compartment and is related to cerebral blood flow and the extraction fraction. The other rate constants are similarly complex, k3 being related to the ligand-receptor association rate and k4 to the dissociation rate.

The mathematical expertise required to implement model-based techniques is not insignificant. From a more practical point of view, the acquisition of ligand plasma concentration data represents a major challenge in clinical departments and for this reason, among others, Laruelle[108] has introduced a method based on constant intravenous infusion of a ligand. Using such techniques it has been shown that SPECT can quantify neuroreceptor density and affinity in the living human.

References

1. Wyper D, Owens J, Brown D, et al. Abnormalities in cerebral muscarinic acetylcholine receptors measured in-vivo with SPECT in patients with Alzheimer's disease. Neurobiol Aging. 1992; 13:S23.
2. Weinberger DR, Gibson R, Coppola R, et al. The distribution of cerebral muscarinic acetylcholine receptors in vivo in patients with dementia. A controlled study with 123IQNB and single photon emission computed tomography. Arch Neurol. 1991;48:169–76.
3. Britton KE. The development of new radiopharmaceuticals. Eur J Nucl Med.1990;16:373–85.
4. Stöcklin G. Tracers for metabolic imaging of brain and heart. Eur J Nucl Med. 1992;19:527–51.

5. Thakur ML. Radiolabelled peptides: Now and the future. Nucl Med Comm. 1995;16:724–32.
6. Winchell HS, Horst WD, Braun L. N-isopropyl [^{123}I] p-iodoamphetamine: Single pass brain uptake and washout; binding to brain synaptosomes and localisation in dog and monkey brain. J Nucl Med. 1980;21:947–52.
7. Moretti JL, Caglar M, Weinmann P. Cerebral perfusion imaging tracers for SPECT: Which one to choose? J Nucl Med. 1995;36:359–63.
8. Kuhl DE, Barrio JR, Huang SC, et al. Quantifying local cerebral blood flow by N-isopropyl-p-[^{123}I]-Iodoamphetamine tomography. J Nucl Med. 1982;23:196–203.
9. Volkert WA, Hoffman TJ, Seger RM, Troutner DE, Holmes RA. ^{99m}Tc-propylene-amine-oxime (^{99m}Tc PnAO); a potential brain radiopharmaceutical. Eur J Nucl Med. 1984;9:511–6.
10. Costa DC, Ell PJ. Brain blood flow in neurology and psychiatry. Edinburgh: Churchill Livingstone; 1991.
11. Ell PJ, Hocknell JML, Jarrit PH, et al. A ^{99m}Tc labelled radiotracer for the investigation of cerebral vascular disease. Nucl Med Comm. 1985;6:437–41.
12. Neirinckx RD, Burke JF, Harrison RC, Forster AM, Andersen AR, Lassen NA. The retention mechanism of technetium-99m HMPAO: intracellular reaction with glutathione. J Cerebr Blood Flow Metab. 1988;8:S4–S12.
13. Costa DC, Lui D, Sinha AK, Jarritt PH, Ell PJ. Intracellular utilisation of Tc-99m d,l HMPAO and T1-201 DDC in rat brain. Nucl Med Comm. 1989;10:459–66.
14. Sharp PF, Smith FW, Gemmell HG, et al. Technetium-99m HMPAO stereoisomers as potential agents for imaging regional cerebral blood flow: human volunteer studies. J Nucl Med. 1986; 27:171–7.
15. Gartshore G, Bannan P, Patterson J, Higley B, McCulloch J. Evaluation of technetium-99m exametazime stabilised with cobalt chloride as a blood flow tracer in focal cerebral ischaemia. Eur J Nucl Med. 1994;21:913–23.
16. Lassen NA, Andersen AR, Friberg L, Paulson OB. The retention of [^{99m}Tc]-D,L-HMPAO in the human brain after intracarotid bolus injection: a kinetic analysis. J Cerebr Blood Flow Metab. 1988;8:S13–S22.
17. Inugami A, Kanno I, Uemura K. Linearization correction of ^{99m}Tc labeled hexamethyl-propylene amine oxime (HMPAO) image in terms of regional CBF distribution: comparison to $C^{15}O_2$ inhalation steady state method measured by positron emission tomography. J Cereb Blood Flow Metab. 1988;8:S52–S60.
18. Lear JL. Quantitative local cerebral blood flow measurements with technetium-99m HMPAO: evaluation using multiple radionuclide digital quantitative autoradiography. J Nucl Med. 1988; 29:1387–92.
19. Bullock R, Patterson J, Park C. Evaluation of 99mTc-hexamethylpropyleneamine oxime cerebral blood flow mapping after acute focal ischemia in rats. Stroke. 1991;22:1284–90.
20. Andersen AR, Friberg HH, Schmidt JF, Hasselbalch SG. Quantitative measurements of cerebral blood flow using SPECT and [^{99m}Tc]-D,L-HMPAO compared to xenon-133. J Cereb Blood Flow Metab. 1988;8:S69–S81.
21. Yonekura Y, Nishizawa S, Mukai T, et al. SPECT with [^{99m}Tc]-D,L-hexamethyl-propylene amine oxime (HMPAO) compared with regional cerebral blood flow measured by PET: effects of linearisation. J Cereb Blood Flow Metab. 1988;8:S82–9.
22. Kotz D. The radiopharmaceutical approval process: a day in the life. J Nucl Med. 1995;36: 15N–30N.
23. Radiopharmacy & Quality Control Pharmacists Subcommittees of the Regional Pharmaceutical Officers Committee. Quality control of radiopharmaceuticals. Nucl Med Comm. 1994;15: 886–9.
24. Rosenthal MS, Cullom J, Hawkins W, Moore SC, Tsui BMW, Yester M. Quantitative SPECT imaging: a review and recommendations by the focus committee of the Society of Nuclear Medicine computer and instrumentation council. J Nucl Med. 1995;36:1489–513.
25. Ljungberg M, Strand SE. Scatter and attenuation and correction in SPECT using density maps and Monte Carlo simulated scatter functions. J Nucl Med. 1990;31:1560–7.
26. Buvat I, Rodriguez-Villafuerte M, Todd-Pokropek A, Benali H, DiPaola R. Comparative assess-

ment of nine scatter correction methods based on spectral analysis using Monte Carlo simulations. J Nucl Med. 1995;36:1476–88.
27. Budinger TF. Advances in emission tomography: quo vadis? J Nucl Med. 1990;31:628–31.
28. Croft BY. Single photon emission computed tomography. Chicago: Year Book Medical Publishers Inc; 1986.
29. Gilland DR. Determination of the optimum filter function for SPECT imaging. J Nucl Med. 1988;29:643–53.
30. Taylor D. Filter choice for reconstruction tomography. Nucl Med Comm. 1994;15:857–9.
31. Kuhl DE, Edwards RQ. Image separation in radioisotope scanning. Radiology. 1963;80:653–62.
32. Meuhllehner G. A tomographic scintillation camera. Phys Med Biol. 1971;16:87–96.
33. Keyes JW. Perspectives on tomography. J Nucl Med. 1982;23:633–40.
34. Links JM. Multidetector single photon emission tomography: are two (or three or four) heads really better than one? Eur J Nucl Med. 1993;20:440–7.
35. Jaszczak RJ, Murphy PH, Huard D. Radionuclide emission computed tomography of the head with Tc-99m and a scintillation camera. J Nucl Med. 1977;18:373–80.
36. Keyes JW, Orlandea N, Heetderks WJ. The humongotron. A scintillation camera transaxial tomography. J Nucl Med. 1977;18:381–7.
37. Mueller SP, Plak JF, Kijewski MF, Holman BL. Collimator selection for SPECT brain imaging: the advantage of high resolution. J Nucl Med. 1986;27:1729–38.
38. Moore SC, Kouris K, Cullum I. Collimator design for single photon emission tomography. Eur J Nucl Med. 1992;19:138–50.
39. Gillen GJ, Gilmore B, Elliott AT. An investigation of the magnitude and causes of count loss artefacts in SPECT imaging. J Nucl Med. 1991;32:1771–6.
40. Jaszczak RJ, Gilland DR, Hanson MW, Jang S, Greer KL, Coleman RE. Fast transmission CT for determining attenuation maps using a collimated line source, rotatable air–copper–lead attenuators and fan bean collimation. J Nucl Med. 1993;34:1577–86.
41. Gardin I, Linhart NC, Petiet A, Bok BD. Dosimetry at the cellular level of kupffer cells after technetium-99m–sulphur colloid injection. J Nucl Med. 1992;33:380–4.
42. Makrigiorgos GM, Adelstein SJ, Kassis AL. Cellular radiation dosimetry and its implications for estimation of radiation risks. JAMA. 1990;264:592–5.
43. Faraggi M, Gardin I, de Labriolle-Vaylet C, Moretti J-L, Bok BD. The influence of tracer localisation on the electron dose rate delivered to the cell nucleus. J Nucl Med. 1994;35:113–9.
44. Kassis AI, Adelstein SJ. Does nonuniformity of dose have implications for radiation protection. J Nucl Med. 1992;33:384–7.
45. International Commission on Radiological Protection. 1990 Recommendations of the International Commission on Radiological Protection. ICRP Publication 60. Oxford: Pergamon Press; 1991.
46. Poston JW. Application of the effective dose equivalent to nuclear medicine patients. J Nucl Med. 1993;34:714–6.
47. Stabin M, Stubbs J, Watson E. Recent controversy in radiation dosimetry. Eur J Nucl Med. 1993;20:371–2.
48. International Commission on Radiological Protection. Addendum 1 to Publication 53 – Radiation dose to patients from radiopharmaceuticals. ICRP Publication 62. Oxford: Pergamon Press; 1993.
49. Boundy KL, Barnden LR, Rowe CC, et al. Human dosimetry and biodistribution of iodine-123-iododexetimide: a SPECT imaging agent for cholinergic muscarinic neuroreceptors. J Nucl Med. 1995;36:1332–8.
50. Hughes JS, O'Riordan MC. Radiation exposure of the UK population – 1993 review. NRPB-R263. London: HMSO; 1993.
51. National Radiological Protection Board. Occupational, public and medical exposure. Documents of the NRPB, Vol. 4 No. 2 London: HMSO; 1993.
52. National Radiological Protection Board. Living with radiation, London: HMSO; 1989.
53. Cohen BL. Test of the linear-no threshold theory of radiation carcinogenesis for inhaled radon decay products. Health Physics. 1995;68:157–74.

54. Macklis RM, Beresford B. Radiation hormesis. J Nucl Med. 1991;32:350–9.
55. National Radiological Protection Board. Risk of radiation-induced cancer at low doses and low dose rates for radiation protection purposes. Documents of the NRPB 1995; 6 No. 1; Chilton: NRPB.
56. Royal College of Physicians. Research on healthy volunteers. London: Royal College of Physicians; 1986.
57. Royal College of Physicians. Research involving patients. London: Royal College of Physicians; 1990.
58. International Commission on Radiological Protection. Radiological Protection in Biomedical Research. ICRP Publication 62. Oxford: Pergamon Press; 1993.
59. World Health Organisation. Use of ionising radiation and radionuclides on human beings for medical research, training and non-medical purposes. Technical Report Series 611, Geneva: WHO; 1977.
60. Mazziotta JC, Phelps ME, Plummer D, Kuhl DE. Quantitation in positron emission computed tomography: 5. Physical–anatomical effects. J Comp Assist Tomog. 1981;5:734–43.
61. Loessner A, Alavi A, Lewandrowski K-U, Mozley D, Souder E, Gur RE. Regional cerebral function determined by FDG-PET in healthy volunteers: Normal patterns and changes with age. J Nucl Med. 1995;36:1141–9.
62. Alavi A, Newberg AB, Souder E. Quantitative analysis of PET and MRI data in normal aging and Alzheimer's disease: atrophy weighted total brain metabolism and absolute brain metabolism as reliable discriminators. J Nucl Med. 1993;34:1681–7.
63. Maziere B, Maziere M. Where have we got to with neuroreceptor mapping of the human brain? Eur J Nucl Med. 1990;16:817–35.
64. Jones EG. Neurotransmitters in the cerebral cortex. J Neurosurg. 1986;65:135–53.
65. Chemical Neuroanatomy. In: Nolte J. The Human Brain. St. Louis, MO: Mosby; 1993: Chapter 17;414–24.
66. Neurotransmitters and neuromodulators. In: Kretschmann H-J, Weinrich W. Cranial neuroimaging and clinical neuroanatomy. Stuttgart: Thieme; 1992: Chapter 7; 342–8.
67. Rinne UK. Brain neurotransmitter receptors in Parkinson's disease. In: Marsden CD, Fahn S, editors. Movement disorders. London: Butterworth Scientific; 1982:59–74.
68. Davies P, Verth AH. Regional distribution of muscarinic acetylcholine receptors in normal and Alzheimer's type dementia brains. Brain Res. 1979;138:389–92.
69. Beer H-F, Blauenstein PA, Hasler PH, et al. In vitro and in vivo evaluation of iodine-123-Ro 16-0154: a new imaging agent for SPECT investigations of benzodiazepine receptors. J Nucl Med. 1990;31:1007–14.
70. Di Chiro G, Brooks RA. PET quantitation: blessing and curse. J Nucl Med. 1988;29:1603–4.
71. Delbeke D, Meyerowitz C, Lapidus RL. Optimal cutoff levels of F-18 fluorodeoxyglucose uptake in the differentiation of low grade from high grade brain tumors with PET. Radiology. 1995;195:47–52.
72. Kahn D, Follett KA, Bushnell DL, et al. Diagnosis of recurrent brain tumor: value of ^{201}Tl SPECT vs ^{18}F-fluorodeoxyglucose PET. Am J Radiol. 1994;163:1459–66.
73. Dierckx RA, Dobbeleir A, Pickut BA, et al. Technetium-99m HMPAO SPECT in acute supratentorial infarction, expressing deficits as millimitre of zero perfusion. Eur J Nucl Med. 1995; 22:427–33.
74. Hanson SK, Grotta JC, Rhoades H, et al. Value of single photon emission computed tomography in acute stroke therapeutic trials. Stroke. 1993;24:1322–9.
75. Johnson KA, Kijewski MF, Becker JA, Garada B, Satlin A, Holman BL. Quantitative brain SPECT in Alzheimer's disease and normal aging. J Nucl Med. 1993;34:2044–8.
76. Front D, Iosilevsky G, Frenkel A. In vivo quantitation using SPECT of radiopharmaceutical uptake by human meningioma. Radiology. 1987;164:93–6.
77. Pretorius PH, van Aswegen A, Lotter MG, Herbst CP, Nel MG, Otto AC. Verification of a varying threshold edge detection SPECT technique for spleen volume: A comparison with computed tomography volumes. J Nucl Med. 1993;34:963–7.
78. Evans AC, Marrett S, Torrescorzo J, Ku S, Collins L. MRI–PET correlation in three dimensions

using a volume of interest (VOI) atlas. J Cerebr Blood Flow. 1991;11(Suppl. 1):A69–A78.
79. Nadeau SE, Couch MW, Devane CL, Shulka SS. Regional analysis of D2 dopamine receptors in Parkinson's disease using SPECT and iodine-123-iodobenzamide. J Nucl Med. 1995;36: 384–93.
80. Crosson B, Williamson DJG, Shukla SS, Honeyman JC, Nadean SE. A technique to localize activation in the human brain with technetium-99m-HMPAO SPECT: A validation using visual stimulation. J Nucl Med. 1994;35:755–63.
81. Mazziotta JC, Pelizzari CC, Chen GT, Bookstein FL, Valentino D. Regions of interest: the relation-ship between structure and function in the brain. J Cereb Blood Flow Metab. 1991; 11:A51–6.
82. Friston KJ, Frith CD, Liddle PF, Frackowiak RSJ. Comparing functional (PET) images: the assessment of significant change. J Cereb Blood Flow Metab. 1991;11:690–9.
83. Thatcher RW, Hallett M, Zeffiro T, John ER, Huerta M, editors. Functional neuroimaging: Technical foundations. London: Academic Press; 1994.
84. Verhoeff NPLG, Kapucu O, Sokole Bussemann E, Van Royen EA, Janssen AGM. Estimation of dopamine D2 receptor-binding potential in the striatum with Iodine-123-IBZM SPECT – Technical and interobserver variability. J Nucl Med. 1993;34:2076–84.
85. Lidow MS, Goldman-Rakic PS, Rakic P, Innis RB. Dopamine D2 receptors in the cerebral cortex: distribution and pharmacological characterisation with ^{3}H raclopride. Proc Natl Acad Sci. 1989;86:6412–6.
86. Shulka SS, Honeyman JC, Crosson B, Williams CM, Nadeau SE. Method for registering brain SPECT and MR images. J Comput Assist Tomogr. 1992;16:966–70.
87. Zhang J, Levesque MF, Wilson CL, et al. Multimodality imaging of brain structures for stereotactic surgery. Radiology. 1990;175:435–41.
88. Golub GH, Van Loan CF. Matrix computations. Baltimore: John Hopkins University Press; 1983.
89. Faber TL, Stokely EM. Orientation of three-dimensional structures in medical images. IEEE Trans Pattern Anal Machine Intell. 1988;626–33.
90. Woods RP, Cherry SR, Mazziotta JC. Rapid automated algorithm for aligning and reslicing PET images. J Comput Assist Tomogr. 1992;14:620–33.
91. Woods RP, Mazziotta JC, Cherry SR. Optimizing activation methods: Tomographic mapping of functional cerebral activity. In: Thatcher RW, Hallett M, Zeffiro T, John ER, Huerta M, editors. Functional neuroimaging, technical foundations. San Diego: Academic Press; 1994:47–57.
92. Andersson JLR. A rapid and accurate method to realign PET scans utilizing image edge information. J Nucl Med. 1995;36:657–69.
93. Levin DN, Pelizzari CA, Chen GTY, Chen CT, Copper MD. Retrospective geometric correlation of MR, CT and PET images. Radiology. 1988;169:817–23.
94. Friston KJ, Frith CD, Liddle PF, Frackowiack RSJ. Plastic transformation of PET images. J Comp Assist Tomogr. 1991;15:634–9.
95. Mazziotta JC, Valentino D, Grafton S, et al. Relating structure to function in vivo with tomographic imaging. In: Exploring brain function with positron tomography. (Ciba Foundation Symposium.) Chichester: J Wiley and Sons; 1991:93–112.
96. Bookstein FL. Landmarks, edges, morphometrics and the brain atlas problem. In: Thatcher RW, Hallett M, Zeffiro T, John ER, Huerta M, editors. Functional neuroimaging: Technical foundations. London: Academic Press; 1994:107–19.
97. Ford I. Some nonontological and functionally unconnected views on current issues in the analysis of PET data sets. J Cerebr Blood Flow Metab. 1995;15:271–7.
98. Thurfjell L, Bohm C, Greitz T, Eriksson L, Ingvar M. Accuracy and precision in image standardisation in intra- and intersubject comparisons. In: Thatcher RW, Hallett M, Zeffiro T, John ER, Huerta M, editors. Functional neuroimaging: Technical foundations. London: Academic Press; 1994:107–19.
99. Wagner HN Jr. A new era of certainty. J Nucl Med. 1995;36(8):13N–28N.
100. Carson RE. The development and application of mathematical models in nuclear medicine. J Nucl Med. 1991;32:2206–8.

101. Huang S, Phelps ME. Principles of tracer kinetic modeling in positron emission tomography and autoradiography. In: Phelps ME, Mazziotta JC, Schelbert H, editors. Positron emission tomography and autoradiography: Principles and applications for the brain and heart. New York: Raven Press; 1986.
102. Laruelle M, van Dyck C, Abi-Dargham A, et al. Compartmental modeling of iodine-123-iodobenzofuran to dopamine D2 receptors in healthy subjects. J Nucl Med. 1994;35:743–54.
103. Laruelle M, Baldwin RM, Rattner Z, et al. SPECT Quantification of ^{123}I-iomazenil binding to benzodiazepine receptors in non-human primates: 1. Kinetic modeling of single bolus experiments. J Cereb Blood Flow Metab. 1994;14:439–52.
104. Kerwin RW, Pilowsky LS. Traditional receptor theory and its application to neuroreceptor measurements in functional imaging. Eur J Nucl Med. 1995;22:699–710.
105. Gjedde A, Wong DF. Modeling neuroreceptor binding of radioligands in vivo. In: Frost JJ, Wagner HN Jr, editors. Quantitative imaging: neuroreceptors, neurotransmitters and enzymes. New York: Raven Press; 1990:51–79.
106. Eckelman WC, editor. Receptor binding radiotracers vol. 2. Florida: CRC Press; 1992.
107. Westera G, Buck A, Burger C, Leenders KL, von Schulthess GK, Schubiger AP. Carbon-11 and iodine-123 iomazenil: a direct PET-SPECT comparison. Eur J Nucl Med. 1996;23:5–12.
108. Laruelle M, AbiDargham A, Al-Tikriti MS, et al. SPECT Quantification of ^{123}I-iomazenil binding to benzodiazepine receptors in non-human primates: 2. Equilibrium analysis of constant infusion experiments and correlation with in vitro parameters. J Cerebr Blood Flow Metab. 1994;14:453–65.

2. SPECT imaging in focal epilepsy

RODERICK DUNCAN

Introduction

Epilepsy is a common disorder, with a prevalence in the western world of approximately 1:200[1]. The great majority of adult epilepsies are focal, most originating in the mesial temporal lobe. About 20% of these epilepsies are not brought under satisfactory control by drug treatment[2] and surgical treatment is being considered with increasing frequency.

The concept that some epilepsies result from disorders of specific parts of the brain and are in principle amenable to surgical treatment has been known and accepted for many years. Since the advent of EEG in the 1930s epileptogenic zones have been defined and localized according to electrical criteria. The development of EEG techniques allowing ictal and intracranial recording has greatly improved effectiveness and accuracy, as judged by the ability of the technique to localize an epileptogenic zone and the confirmation of that localization by a good surgical result.

Some of the EEG techniques used in presurgical localization are invasive, expensive or both. In particular, the stereotaxic placement of intracranial electrodes requires a considerable commitment to surgery, both financial on the part of the medical care agencies, and personal in terms of the risk and time the patient is prepared to take to rid himself of his seizures. This kind of consideration has limited the availability of surgical treatment and, at the same time, provided the impetus to develop alternative methods of localization. Advances in anatomical imaging, principally MRI, have allowed the detection of more and more subtle anatomical lesions, but epilepsy and seizures are primarily disturbances of brain function.

In the last two decades, the two main functional imaging techniques, PET and SPECT, have been used extensively to detect focal disturbances in brain function in patients with focal epilepsies, with PET being used mainly to image local cerebral metabolic rate for glucose (lCMRGlc)[3] and SPECT to image regional cerebral blood flow[4]. Used with the tracers HIPDM and HMPAO, SPECT currently provides the only practical means of imaging brain function during or soon after seizures, and, of all functional imaging techniques, has proved most useful as a localizing investigation. Much of its utility, especially in the ictal and postictal periods, results from the spatial and temporal characteristics of the rCBF SPECT method. Intracranial EEG recordings give localizing information which is continuous in time, but which of necessity relates only to a preselected series

R. Duncan (ed): SPECT Imaging of the Brain, 43–68.

of points in the brain. HMPAO SPECT shows the whole brain, and the image obtained provides localizing information which relates essentially to an epoch lasting 40 s (the period of uptake of the tracer), thus giving a completely different perspective from that provided by EEG techniques. The bulk of this chapter will therefore deal with the current use and continuing development of SPECT to image regional cerebral blood flow in focal epilepsies.

Ligands for neurotransmitter receptors for use with SPECT have become available in recent years, most notably in the field of epilepsy, for central type benzodiazepine receptors. Others are being developed, and an account will be given of this field from the clinical and research points of view.

rCBF and the pathophysiology of seizures and epilepsy

Why study cerebral blood flow in epilepsy? Whether we are speaking of interictal or perictal cerebral blood flow, the answer lies in the link between the electrical activity of the brain, the metabolic activity required to provide the energy for that electrical activity, and the blood flow required to deliver the substrates to provide the energy. Essentially, local changes in blood flow reflect local changes in the level of brain function. In the interictal period, hypoperfusion may simply represent the presence of an anatomical lesion, or may indicate more widespread hypofunction due to neuronal loss, inhibition or other factors. During and after seizures hyperperfusion, hypoperfusion, or a combination of both may be seen, directly reflecting ictal and postictal changes in the level of neuronal activity in structures involved in the origin and propagation of seizure discharges.

The existence of changes in cerebral blood flow during seizures has been known for a century, originally from direct observation of the cerebral cortex. Some of these early observations were remarkable for their quality and detail, and still provide some information not readily available to us today.

Early studies of cerebral blood flow in epilepsy

Sir Victor Horsley described focal ictal hyperfusion in 1892[5], but Penfield[6–8] was the first to make a systematic study of vascular changes associated with seizures. He pioneered the use of resective surgery for focal epilepsy and used intraoperative stimulation of the cerebral cortex in the area of the focus to induce seizures. In this way, he was able to make direct observations of changes in the pial blood vessels. Because intraoperative induction of seizures is no longer carried out, no such observations have been made for many years. He described a total of 49 examples of seizures on conscious patients whose cerebral cortex was open to direct observation on the operating table. He noted vascular changes in most of these cases, including ictal hyperperfusion and postictal hypoperfusion. Using the heated thermocouple method, he and later workers were able to demonstrate the time course of ictal changes in cerebral blood flow[6–8], showing that rises in

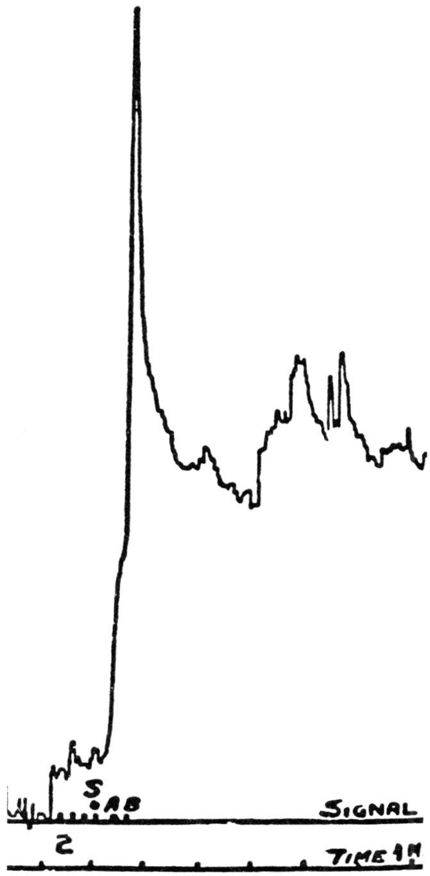

Figure 2.1. Time course of blood flow changes during a seizure, recorded by Gibbs et al. Vertical axis: temperature of a heated thermocouple placed near the epileptogenic zone; horizontal axis: time (min). The seizure starts at S, terminating at B.

flow began in the cortex adjacent to the focus 4–10 s after the first convulsive movement and continued for some minutes after the convulsions had ceased (Figure 2.1). Accompanying changes in blood flow were seen in the basal ganglia: these were of greater magnitude than those found in the cortex. Thermocouples placed distant from the focus showed either no increase in blood flow or a preliminary decrease.

Penfield thought that these changes in blood flow were consequent on changes in metabolic rate brought about by neuronal activity. His observations and those of Gibbs et al. contain the major elements of the perictal changes in blood flow which have been demonstrated in recent years using HMPAO SPECT, and provide information on the time course of such changes not readily available to us today. Dymond and Crandall[9] made similar measurements of brain temperature in patients during temporal lobe seizures, the results of which were consistent with those of Penfield and Gibbs.

Penfield eventually thought that the changes in blood flow he observed were simply consequent on the metabolic demand of neuronal activity. This is likely to be the case, although our knowledge of flow/metabolism coupling during and soon after seizures remains incomplete.

Relationship between cerebral blood flow and cerebral metabolic rate

The maintenance of ionic gradients, and therefore the generation of electrical activity, is probably responsible for 80% of the energy requirement of the brain in the resting state[10]. Metabolism and flow are closely coupled in the brain, on a moment-to-moment and on a local basis. The mechanism of the coupling remains a subject of debate, but it is likely that vasodilatation is mediated by a product of neuronal metabolism, as originally postulated by Roy and Sherrington in 1890[11]. While changes in blood flow are proportional to changes in metabolic rate, the slope of the relationship may vary, and in some physiological circumstances blood flow may increase proportionally more than metabolic rate[12].

The available evidence suggests that this normal close coupling between blood flow and metabolism holds true in epilepsy, with the possible exception of the postictal period. Plum et al.[13] and Posner et al.[14] showed that the increase in total CBF during seizures exceeded that required for adequate oxygenation of the brain as a whole. They also showed that the venous oxygen tension rose after generalized seizures, giving a counterpart to Penfield's observations of saturated venous blood draining from areas of brain involved in focal seizures, and showing that blood flow remained high after cessation of seizure activity. Although some animal studies have shown local mismatches between perfusion and metabolism during seizures[15], methodological problems make the significance of this uncertain. PET studies in human status epilepticus suggest that blood flow is adequate for metabolic demand[16], although the spatial resolution of in vivo imaging techniques is not yet sufficient to detect very localized mismatches.

Interictal cerebral metabolic rate in epilepsy

Although this chapter is concerned with regional cerebral blood flow, it is worthwhile considering studies of local cerebral metabolic rate, since the two parameters are closely linked. Contrasts as well as comparisons can be made. Almost all studies have been of temporal lobe epilepsy, as this is the most common form of focal epilepsy in adults and that most easily treated surgically.

Local cerebral metabolic rate has been studied using PET, largely with ^{18}FDG[17–21]. The most common finding is hypometabolism of the temporal lobe containing a mesial focus, with some patients showing more widespread or bilateral hypometabolism. The proportion of patients in whom an abnormality is found appears to be higher when FDG PET is used than when SPECT is used to image rCBF (in most series approximately 80% compared with 50–80%; see below). One study[22] found a majority of patients with mesial temporal foci (13/17) in whom hypometabolism was greater in degree in the lateral temporal cortex. As might be expected, changes in the basal ganglia and thalamus mirror cortical changes[23].

The site of FDG PET abnormalities correlates well with EEG localization of the epileptic focus, although the extent of hypometabolism tends to be much greater than the epileptogenic area or any associated lesion[24]. Theodore et al.[25] found that the duration of epilepsy in a group of 20 patients correlated inversely with their mean lCMRGlc, but that there was no correlation with the age of the patient or the age at onset of epilepsy. Anticonvulsant drugs such as barbiturate and phenytoin tend to reduce lCMRGlc, as does carbamazepine to a lesser degree[17].

Interictal studies of rCBF

Over the last decade, the principal method used to study rCBF has been SPECT using either rotating gamma cameras or dedicated head imagers, with xenon, IMP, HIPDM or latterly HMPAO[26,27] as tracers. PET, using the oxygen inhalation or intravenous ammonia techniques, and the XeCT method have also been used. SPECT has been carried out on a large number of patients with focal epilepsy, in most cases originating in the mesial temporal lobe structures.

Hypoperfusion due to anatomic lesions

Where there is an anatomical lesion local hypoperfusion is often due to the lesion itself. This is usually easy to recognize, even before the anatomical images are seen, as the area of hypoperfusion usually has well demarcated boundaries and manifests simply as an 'interruption' of the relatively high flow in the cortex (Figure 2.2). Small lesions, especially those situated in white matter, may produce no detectable change in rCBF, depending on the spatial resolution of the imager. Some lesions are associated with areas of hypoperfusion much larger than the lesion itself. This is also true for abnormalities found on pathological examination of resected surgical specimens[28,29]. Focal or hemispheric atrophy is not uncommon in epilepsy and may produce the appearance of hypoperfusion simply by reducing the volume of cortex, or as a result of glial change.

Clearly, any functional image must be interpreted in the light of the anatomy. This may seem an obvious statement, but hypoperfusion which has an anatomical explanation has no significance other than the presence of the lesion, and will be found whether the lesion causes seizures or not. The finding nonetheless is of considerable importance as a reference for ictal or postictal studies. Epileptogenic tissue at the margin of a lesion may be hypoperfused interictally. During the seizure, perfusion may increase to levels not very different from those of surrounding tissue. To appreciate the change that has taken place (or to be sure that a change has not taken place) detailed comparison of the images will be required. It is

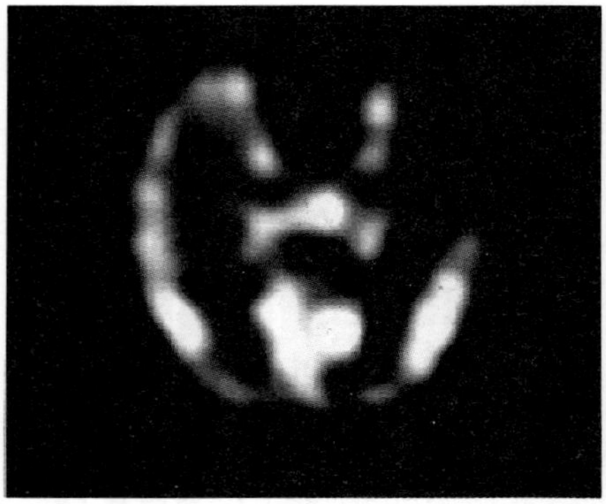

Figure 2.2. HMPAO SPECT: axial slice in the long axis of the temporal lobe of a patient with complex partial seizures and a tumour in the left anterolateral temporal lobe. The tumour was evident on MRI but CT was normal. HMPAO SPECT shows a sharply delineated gap in the neocortex and the pole of the left temporal lobe.

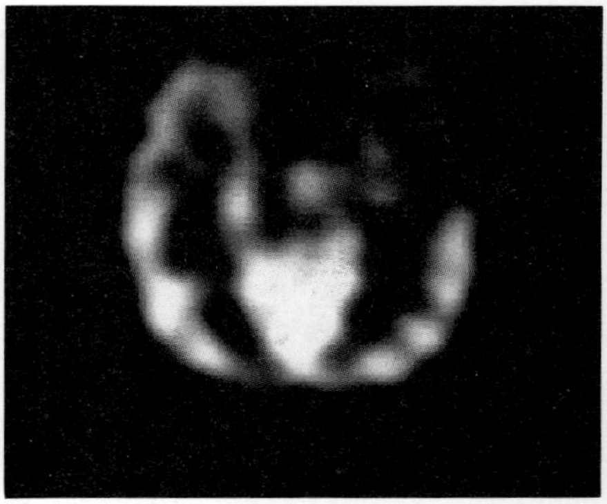

Figure 2.3. Interictal HMPAO SPECT: axial slice in the long axis of the temporal lobe of a patient with complex partial seizures of left temporal origin, showing hypoperfusion of the left temporal lobe, more marked in the lateral cortex (mesial asymmetry index 9%, lateral asymmetry index 17%).

always important to assess ictal and postictal changes with reference to interictal images, particularly where the pattern of rCBF is abnormal in the first place.

Hypoperfusion not associated with gross anatomical lesions

Abnormalities of rCBF have been found in between approximately 50% and 80% of patients without gross anatomical lesions[17,30–36]. The reason for this variability is unclear, but in some instances no normal data have been presented, and in others exclusion of anatomical lesions has been based on low resolution structural imaging. Some series are relatively small, making it possible that differences in patient selection have a major effect. For example, the age at onset of epilepsy appears to have an important effect on the likelihood of finding abnormal rCBF (see below), and such data are not usually specified. Differences in reporting practice may also produce large differences in abnormality rate, but the lack of numerical data in most studies makes meaningful comparison difficult.

In the majority of patients in whom rCBF is abnormal, the abnormality found is focal hypoperfusion, usually involving all or most of the temporal lobe (Figure 2.3). The hypoperfusion may be predominantly mesial or lateral, more often the former. In a series of 41 patients with confirmed mesial temporal foci 24 had predominantly lateral hypoperfusion[43]. In a minority of patients (e.g. one of the same series of 41 patients) hypoperfusion may affect only, or at least very predominantly, one part of the temporal lobe, in almost all cases the mesial temporal lobe (Figure 2.4), remembering that these figures exclude patients with detectable anatomical lesions. More frequently, hypoperfusion extends outside the temporal lobe, most commonly into the ipsilateral orbitofrontal cortex (Figure

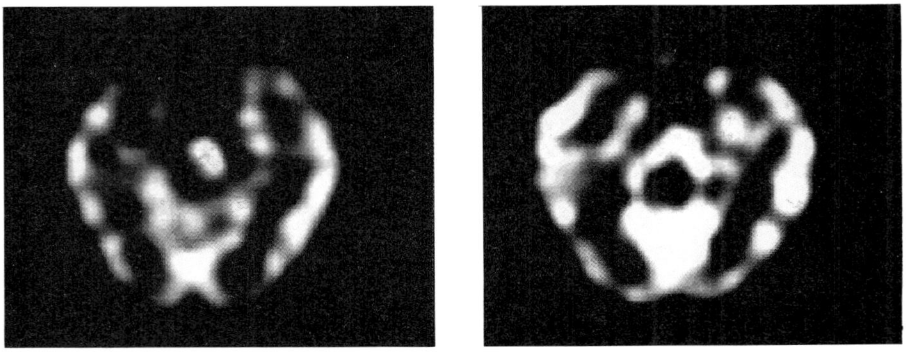

(a) (b)

Figure 2.4. (a) Interictal HMPAO SPECT: axial slice in the long axis of the temporal lobe of a patient with complex partial seizures of left temporal origin, showing hypoperfusion of the left mesial temporal cortex and pole (mesial asymmetry index – 15%, lateral asymmetry index 4%). (b) Repeat interictal image in the same patient, showing symmetric flow. Postictal HMPAO SPECT and ictal EEG recordings showed right mesial temporal origin for the seizures.

2.5), but sometimes also involving the whole frontal lobe or hemisphere. Bilateral temporal hypoperfusion is occasionally seen.

'Interictal hyperperfusion'

Some series[33,37] have found that a small proportion of patients show focal hyperperfusion interictally, or a combination of hypoperfusion and hyperperfusion in different areas or even contralaterally. The hyperperfusion may be situated in areas remote from the epileptogenic area. The reason for this finding is unclear. One patient from the Glasgow postictal series[38] was injected 1.5 h after a series of seizures and showed gross hyperperfusion (Figure 2.6), and we have since seen several instances of hyperperfusion persisting for over 24 h postictally. Indeed, ictal hyperperfusion was noted in one case to persist for several days postictally[39]. This underlines the need for interictal examinations to be performed as long as possible after the last seizure.

Whatever the cause of interictal hyperperfusion, it is difficult to believe that it represents increased neuronal activity when both behaviour and EEG are normal, particularly since interictal hypermetabolism is rarely found in PET studies. Recent data suggest the possibility that HMPAO uptake overestimates blood flow in areas of reperfusion after cerebral infarction[40]. It is possible that this happens in other circumstances, such as following repeated or lengthy seizures, and that these areas of high HMPAO uptake do not represent high flow at all.

In practical terms, it would probably be unwise at present to ascribe any localizing significance to focal hyperperfusion unless it is accompanied by either clinical or EEG evidence of seizure.

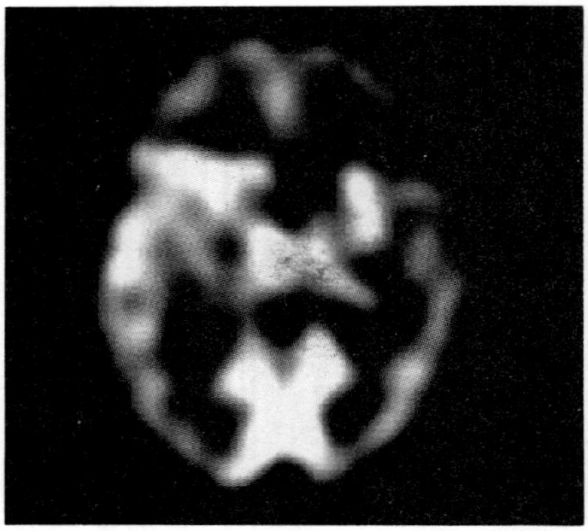

Figure 2.5. Interictal HMPAO SPECT: axial slice in the orbitomeatal plane showing left temporal and frontal hypoperfusion in a patient with complex partial seizures of left temporal origin.

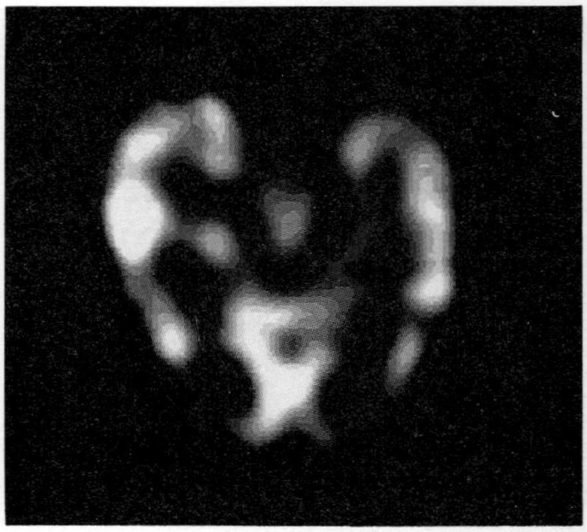

Figure 2.6. Interictal HMPAO SPECT: axial slices in the long axis of the temporal lobe showing a focus of hyperperfusion in the lateral cortex of the right temporal lobe. The asymmetry in flow was 16%. Postictal SPECT and ictal EEG recordings showed left temporal origin for seizures, and left temporal lobectomy resulted in freedom from seizures.

Causes and significance of interictal hypoperfusion in temporal lobe epilepsy

The cause of interictal hypoperfusion (and hypometabolism) is unknown, but its associated factors suggest some possibilities. Correlations between abnormalities of rCBF and clinical factors vary, but correlations have been found with seizure frequency[41], length of seizure history[42], and with age at onset of epilepsy. The last factor appears to be the most important; in a series of 89 patients, Duncan et al.[34] found a striking relationship between age at onset of epilepsy and abnormal rCBF, with patients with early onset disease being much more likely to have abnormal rCBF (Figure 2.7). This finding has been confirmed in a similarly large ($n=81$) subsequent series[43]. The proportion of patients with abnormal rCBF as determined by blinded visual analysis of images decreased abruptly after age at onset of epilepsy at 5 years. This suggests that age at onset of epilepsy is an important determinant (or associated factor) of the occurrence of temporal hypoperfusion.

Neuronal loss occurs in the hippocampus of patients with mesial temporal epilepsy[44,45], a process which is thought to relate to the production of excess excitotoxic neurotransmitters such as glutamic acid[46] during seizures. Neuronal loss is commonly found in the hippocampus of patients with temporal hypometabolism[20], and this may apply equally to hypoperfusion, though no study has yet been carried out. Hippocampal neuronal loss might therefore contribute to low metabolism and blood flow not only locally but also more distantly in the lateral cortex due to deafferentation effects.

Anomalous localization of memory and other functions to the side contra-

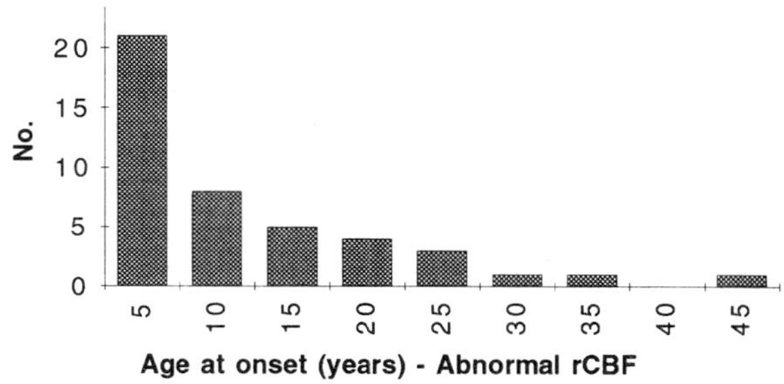

Figure 2.7. Histogram of age at onset of epilepsy (horizontal axis) vs. proportion of patients with abnormal rCBF (vertical axis) in a population of 89 patients with complex partial seizures of temporal lobe origin. Abnormal rCBF was more common in patients with epilepsies of early onset. There was no clear relationship with patient age or length of history of epilepsy.

lateral to the focus may occur in temporal lobe epilepsy of early onset[47]. This means that the temporal lobe that is the site of the focus does not function, and this effect might contribute to hypoperfusion and hypometabolism. This effect would also be consistent with the correlation found between hypoperfusion and epilepsy of early onset.

Interictal and ictal epileptiform activity has been associated with neuronal inhibition in surrounding tissues (surround inhibition[48]) and ictal discharges are associated with hypometabolism and hypoperfusion in surrounding tissues, in man and in animals (see below). However, in two studies[21,49] no correlation was found between interictal hypometabolism and spiking on EEG.

Whatever its cause(s), it is likely that interictal hypoperfusion represents a functional change that is the consequence of seizures, rather than being an intrinsic part of the epileptogenic process.

Problems of localization of the epileptogenic zone using interictal rCBF SPECT

The correlation between the location of abnormalities of rCBF and EEG localization has been variable. In general, studies showing a high proportion of patients with localizing abnormalities[32,50] tend to have a high proportion (up to 40%) of patients in whom SPECT localization disagrees with EEG, while studies showing a low abnormality rate tend to have fewer disagreements[36,38]. In one series of 48 patients with mesial temporal epilepsies with good conventional localizing, only 17 of these patients were judged by blinded visual analysis to have unilateral temporal hypoperfusion, in 16 cases on the side of the focus. The remaining patient had unequivocal hypoperfusion restricted to the mesial temporal cortex on the right, but seizures originating from the left mesial temporal lobe (demonstrated both by postictal SPECT and videotelemetry with foramen ovale electrodes).

Interestingly, repeat interictal SPECT in this patient showed resolution of the hypoperfusion (Figure 2.8). One study reported a localization rate of 73% with only one out of 34 patients falsely localized[51], although this surprisingly high sensitivity was achieved in a population in which the examination was repeated in patients whose rCBF was initially normal (note also that the tracer HIPDM was used, rather than HMPAO).

Differences in reporting practices, imagers, tracers and patient populations might all be responsible for variable experiences of interictal SPECT as a localizing investigation. It is nonetheless agreed that there is some risk of false localization. Whether the difference in sensitivity between PET and SPECT reflects a difference in the techniques used rather than in the parameters studied is unclear. Kuhl et al.[18] studied a group of patients using ^{18}FDG to image LCMR and ^{13}NH3 to image rCBF and found similar results with both; Franck et al.[16] found no variation in oxygen extraction ratio (OER) in hypometabolic zones in patients with focal epilepsy. By contrast Stefan et al.[29] used HMPAO SPECT and FDG PET to image rCBF and LCMRGlc in a group of patients and found that PET detected abnormalities that SPECT did not, albeit the SPECT system used was of relatively low resolution. Pathological variability in the slope of the relationship between flow and metabolism (see above) might account for the lower sensitivity of SPECT; i.e. reductions in rCBF might be proportionally less than the accompanying reductions in metabolic rate, and more often fall within normal limits. The fact that HMPAO SPECT appears to be less specific as well as less sensitive is difficult to explain.

Conservative reporting probably reduces the false localization rate of interictal HMPAO SPECT but, until a method for directly and specifically detecting the abnormal excitability which gives rise to seizures is developed, confident localization of an epileptogenic zone still rests on the demonstration of a localized physiological change which takes place in temporal association with the clinical event. In the great majority of patients, we would therefore see the main clinical function of interictal rCBF studies as providing a baseline for the assessment of ictal and postictal changes.

Interictal rCBF in extratemporal epilepsies

There has been limited study of extratemporal epilepsies. One Glasgow series[34] includes only 4/36 patients with extratemporal epilepsies who had abnormal interictal rCBF not associated with a structural abnormality visible on MRI images. In all four cases the abnormalities were well delineated with relatively sharp boundaries, leading to the suspicion that a structural abnormality too subtle for MRI to detect was in fact present, though no pathological confirmation is yet available. One other series has produced a similar result[51].

Although information remains scant, therefore, it appears probable that the localizing sensitivity of interictal HMPAO SPECT will turn out to be even lower in extratemporal than in temporal epilepsies.

Perictal rCBF in focal seizures

Conventionally, the most reliable methods for localizing epileptogenic zones rely on detecting a localized physiological change in association with a clinical event. In practice, this involves recording EEG changes, from the scalp or from within the cranium, during a seizure. In a similar way, perictal SPECT can be used to detect local changes in rCBF during or soon after seizures. However, SPECT using HMPAO and similar tracers has different temporal and spatial characteristics compared with EEG techniques, principally in terms of sampling error, and this must constantly be borne in mind when interpreting results.

In recording a seizure using EEG the sampling error is in space. Recordings are taken from a series of points, on the scalp, on the surface of the brain or within the brain. While nothing is seen of what is happening distant from those points, the recording is continuous in time. The image of rCBF obtained using HMPAO SPECT is complete in space, and is of the whole brain. It is, however, limited in time: HMPAO uptake can only begin when the substance reaches the brain. Before that can happen, the seizure has to be recognized, the injection mixed and given, and the HMPAO has to reach the brain. The transit time from arm to brain is approximately 15 s, and the uptake period is approximately 40 s[53]. The image acquired is one of rCBF approximately averaged over this period, and will reflect any changes of sufficient magnitude, extent and duration which have taken place during this time.

It is often assumed that the time course of ictal changes in rCBF is the same as that of EEG changes. This is not always the case, the obvious example being the persistence of mesial temporal hyperperfusion for several minutes after the end of mesial temporal lobe seizures. Experience in Glasgow suggests that ictal hyperperfusion also persists after shorter seizures too, but for a shorter time. Changes may be seen when HMPAO has been injected at the end of a seizure lasting 20 s, but with seizures shorter than this, the injection has to be given at or near the seizure onset, something which is only possible if the seizure can be provoked or if it occurs very frequently, when the injection can be premixed and the syringe connected to the cannula ready to inject.

This brings us to some practical considerations with respect to the stability of blood flow tracers after mixing. Once mixed with ^{99m}Tc, HMPAO is useable for only 30 min. ^{99m}TcECD (Bicisate–Neurolite, Dupont Pharma[54,55]) is stable for several hours, and HMPAO can be rendered stable by mixing with cobalt chloride, although the latter product is unlicensed. rCBF SPECT is most widely used to localize mesial temporal lobe seizures. These seizures tend to be relatively long (e.g. ictal phase 1–2 min) but relatively infrequently (typically 1–5/week in patients in hospital with medication reduced or stopped). Mixing of HMPAO and ^{99m}Tc pertechnetate can be carried out in approximately 10–15 s. This period of time is seldom significant with respect to the long time scale of the seizure and little is gained by using a premixed stable tracer. Moreover, premixed doses may be wasted unless they can be used for other patients before the compound degrades or the ^{99m}Tc decays. In patients in whom seizures are short and frequent,

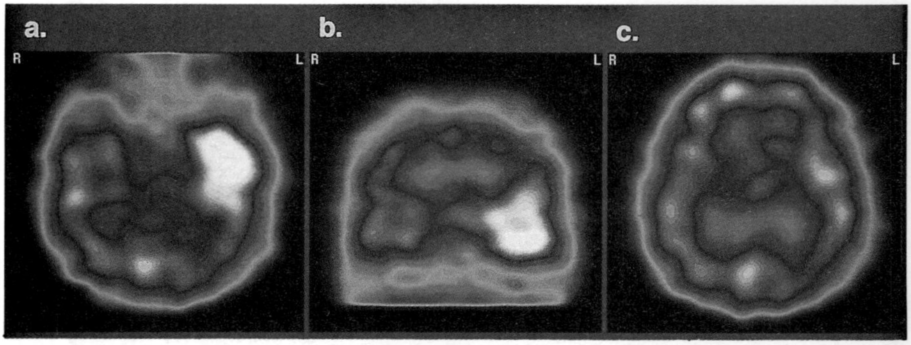

Figure 2.8. Ictal HMPAO SPECT injection during a complex partial seizure of left mesial temporal origin. All three images are from the same dataset. (a) Axial slice in the long axis of the temporal lobe showing marked hyperperfusion of the whole left temporal lobe. (b) Coronal slice through the mid-temporal lobe, showing hyperperfusion of the whole left temporal lobe, extending into the adjacent frontal cortex. More superiorly, there is mild hypoperfusion on the left frontal cortex. (c) Axial slice in the orbitomeatal plane, showing hyperperfusion of left basal ganglia and thalamus. (For colour version of this figure see page 180.)

such as may be encountered in the frontal lobe, there is a high likelihood of a seizure occurring on a given day. The few seconds spent mixing the compound become more important in the context of the short time frame of these seizures, and stable rCBF tracers will certainly have advantages in this situation. Stable compounds may allow the use of rCBF SPECT where local regulations demand that the mixing of isotopes and markers take place in the department of nuclear medicine.

Perictal rCBF in mesial temporal lobe seizures

There is considerable experience of rCBF changes associated with mesial temporal lobe seizures. Ictal studies usually show gross hyperperfusion of the whole temporal lobe (Figure 2.8). This may be accompanied by hypoperfusion of the surrounding cortical tissues, most often of the adjacent orbital cortex, the whole ipsilateral frontal lobe or the whole ipsilateral hemisphere[38,56–60].

Ictal hyperperfusion of the lateral temporal cortex is presumably due to activation as part of propagation of the seizure discharge. Other structures are commonly hyperperfused, including the ipsilateral orbital cortex, the ipsilateral motor cortex, the contralateral temporal lobe and the basal ganglia (Figure 2.8a–c, see also colour section). The hypoperfusion seen in cortical structures ipsilateral to the mesial temporal lobe seizure may be related to inhibition of surrounding cortex. It certainly has its counterpart in animal models of epilepsy[61] and has been related to electrical inhibition in this situation[48], where it has been termed 'surround inhibition'.

In the immediate postictal period (0–4 min) the lateral temporal cortex rapidly

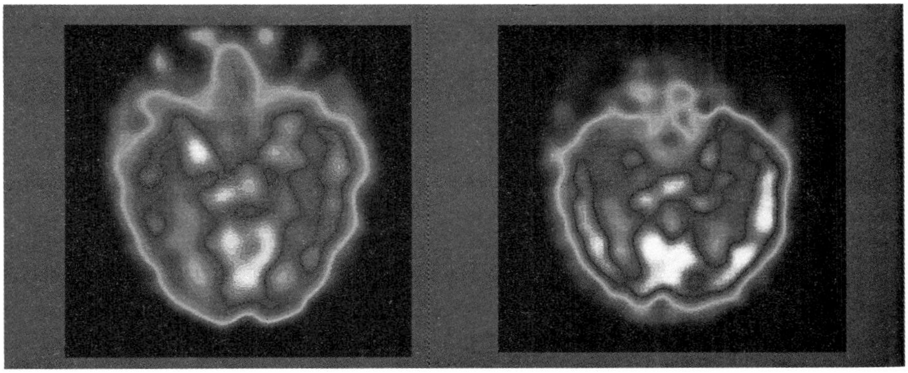

(a) (b)

Figure 2.9. (a) HMPAO injection 1 min after the end of a complex partial seizure of right mesial temporal origin. Axial slices in the long axis of the temporal lobe, showing hypoperfusion of the right lateral temporal cortex. This extended to involve the whole of the right hemisphere, with the exception of the right mesial temporal cortex, where blood flow remains high. (b) Right mesial temporal seizure in a different patient, in whom HMPAO was injected 6 min after the end of a right mesial temporal lobe seizure. This shows hypoperfusion of both mesial and lateral temporal cortex on the right. (For colour version of this figure see page 181.)

becomes hypoperfused (Figure 2.9, see also colour section), often to a profound degree. Again, this often extends into the ipsilateral frontal lobe, and may involve the whole hemisphere[28,38,62]. Areas of hypoperfusion occasionally appear in the contralateral hemisphere and may cause difficulty in interpretation, particularly if the exact temporal relationship of the injection to the seizure discharge is not known. During this time, the mesial temporal cortex remains hyperperfused, only becoming isoperfused relative to the contralateral mesial temporal cortex later in the postictal period (4–15 min). It too then becomes hypoperfused, but even relatively late in the postictal period there is usually some preservation of mesial perfusion levels relative to the lateral cortex.

The sensitivity of perictal SPECT in mesial temporal epilepsies appears to depend to some extent on the timing of the injection. In mesial temporal lobe seizures, true ictal injections appear to show changes in 100% of cases[59], although the appearance of bilateral temporal lobe changes does occasionally prevent confident lateralization. Postictal injections have a sensitivity of 70–90%, depending on the time between the end of the seizure and the injection[38,62]. One major problem with postictal rCBF SPECT in mesial temporal lobe seizures is that up to 20% of patients with unilateral seizure onset show bilateral and symmetrical changes. This provides supporting evidence for a mesial temporal lobe origin for the seizure, but none for its lateralization.

The specificity of ictal rCBF SPECT in mesial temporal lobe seizures seems to be close to 100%. None of the major centres carrying out this kind of work has reported a false localizing study, and we have to date encountered none. There have been reports of occasional false localizing postictal rCBF SPECT studies[59]

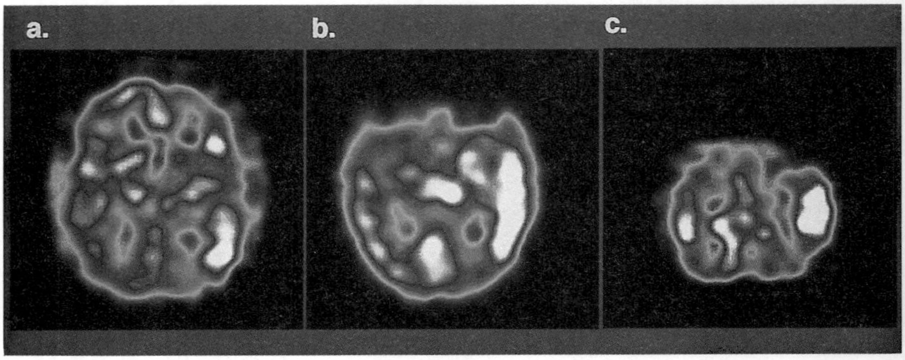

Figure 2.10. HMPAO SPECT. Injection during a complex partial seizure originating in area of the left temporo-parieto-occipital junction (TPOJ). All three images are from the same dataset. (a) Axial slice showing intense hyperperfusion in the area of the TPOJ, with a small focus of hyperperfusion in the ipsilateral inferolateral frontal cortex, with hypoperfusion of the rest of the ipsilateral frontal lobe. (b) Axial slice in the long axis of the temporal lobe showing hyperperfusion of the whole temporal lobe. Note that, unlike the pattern seen in mesial temporal lobe seizures, hyperperfusion is clearly more marked in the lateral as opposed to the mesial temporal cortex. (c) Posterior coronal slice, showing hyperperfusion of the TPOJ, with intense surrounding hypoperfusion. There is a small area of hyperperfusion in the contralateral parietal cortex. (For colour version of this figure see page 181.)

in which changes are subtle. There is no example of this in the Glasgow series at the time of writing.

The timing of the injection notwithstanding, it is crucial to remember that whatever localizing information is obtained using rCBF SPECT, it pertains *only* to that seizure during which the patient was injected (a minority of patients have bilateral or even multiple foci).

rCBF changes in seizures originating outside the mesial temporal lobe

There are relatively few data in the literature regarding rCBF changes in seizures originating outside the mesial temporal structures. Nonetheless, the studies that have been published show patterns of perfusion clearly different from those seen during mesial temporal seizures[52,63–65].

Seizures originating at the temporo-parieto-occipital junction

The following is based on the author's data[63,66], acquired in Glasgow and Rennes, France. An ictal rCBF pattern is illustrated in Figure 2.10 (see also colour section). These seizures are associated with ictal perfusion increases in the area of seizure onset. There is also variable hyperperfusion of the anterolateral temporal lobe structures, mesial temporal structures being hyperperfused to a lesser degree, or even hypoperfused. There may be a small area of hyperperfusion in the contralateral parietal lobe, which appears to correlate with bilateralization of the seizure discharge. Some patients have only isolated hyperperfusion of the TPOJ area.

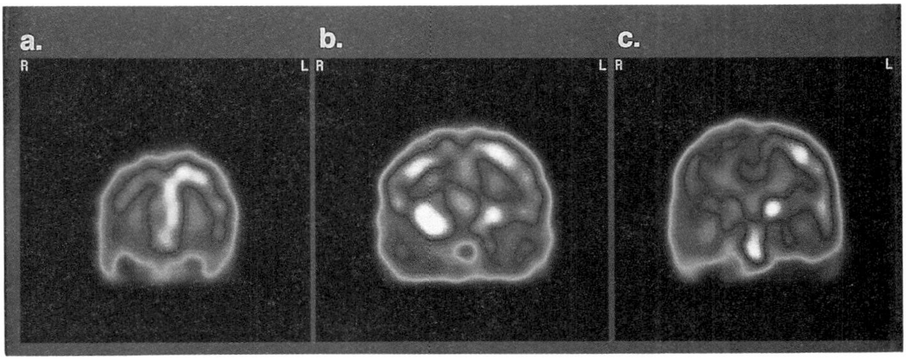

Figure 2.11. HMPAO injection during a complex partial seizure of left mesial frontal origin (MRI and pathological examination indicated focal cortical dysplasia at this site). (a) Coronal slice through the anterior frontal lobe, showing left mesial frontal hyperperfusion. (b) Coronal slice through the basal ganglia, frontal and temporal lobes, showing hyperperfusion of the right basal ganglia, with relative hypoperfusion of the temporal lobes. (c) More posterior coronal slice showing hyperperfusion of the left basal ganglia extending down into the brain stem. The posterior part of the right cortical mantle is hypoperfused. (For colour version of this figure see page 182.)

Frontal lobe seizures

Frontal lobe seizures may present considerable localization difficulties. The frontal lobe is a large structure in which seizure discharges often spread and bilateralize very rapidly, making invasive EEG monitoring necessary in many cases. Although seizures may be very frequent, their brevity poses problems for rCBF SPECT, and many seizures are simply too short to be captured by this method.

Seizures originating in the frontal lobes appear to be associated with a variety of ictal changes, apparently according to the site of origin within the frontal lobe[52,64–68]. It is difficult to make generalizations from the small amount of data available, but they all have in common perfusion abnormalities which are distinct from those seen in mesial temporal lobe seizures, and which appear to be consistent with the site of origin of the seizure within the frontal lobe, where this is known. Associated hyperperfusion of subcortical structures such as the basal ganglia, thalamus and cerebellum seems to be common. One further interesting characteristic of frontal seizures is that ictal hyperperfusion may involve the hemisphere contralateral to the focus (author's data; Figure 2.11, see also colour section), although ipsilateral hypoperfusion is also seen. From this it can be seen that perfusion patterns in frontal lobe seizures may be complex, and that images require very careful interpretation. In particular, it may be dangerous to draw any conclusions from images which seem to show hypoperfusion alone, since the imager may not have detected a small or mildly abnormal area of hyperperfusion in the contralateral frontal lobe.

Parietal lobe seizures

Little has been published, but pure parietal lobe seizures seem to be associated simply with localized hyperperfusion in the parietal lobe, if any changes are seen at all[67] (Figure 2.12, see also colour section).

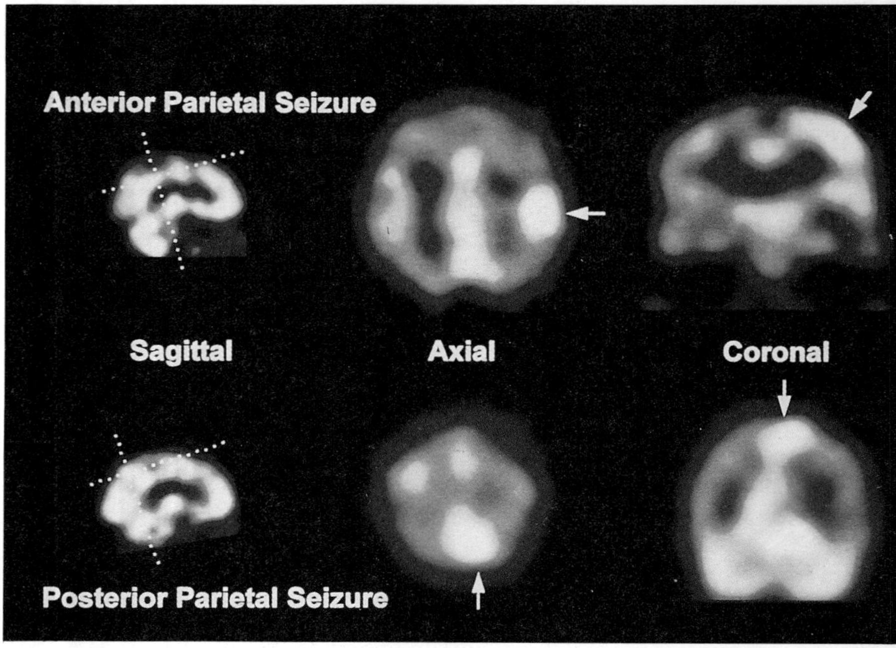

Figure 2.12. Parietal lobe epilepsy. The upper images show a seizure arising from the anterior parietal lobe in a patient whose seizure comprised tingling in the right hand followed by asymmetrical bilateral dystonic posturing. There is hyperperfusion seen in the anterior parietal area. The lower images show a posterior parietal seizure, manifested by an experiential aura followed by a psychoparetic complex partial seizure. The hyperperfusion is more posterior. (Illustration kindly supplied by Dr S. Berkovic, Austin Hospital, Melbourne, Australia). (For colour version of this figure see page 182.)

Occipital lobe seizures

There are no series to date in the literature, and the author's experience is limited to a few cases, which have shown occipital hyperperfusion. An example is shown in Figure 2.13 (see also colour section).

Interpretation of perictal rCBF images

The interpretation of images is dealt with more fully in Chapter 1, but image interpretation presents particular problems in the context of perictal rCBF changes, which may be complex and can only be fully assessed by comparison of perictal and interictal images. Where the only change seen is a localized area of hyperperfusion interpretation is relatively straightforward, but it may not simply be a case of deciding that the seizure originates at the 'bright spot' on the scan. For this reason, the process of interpretation may be conceptually divided into 2 stages:

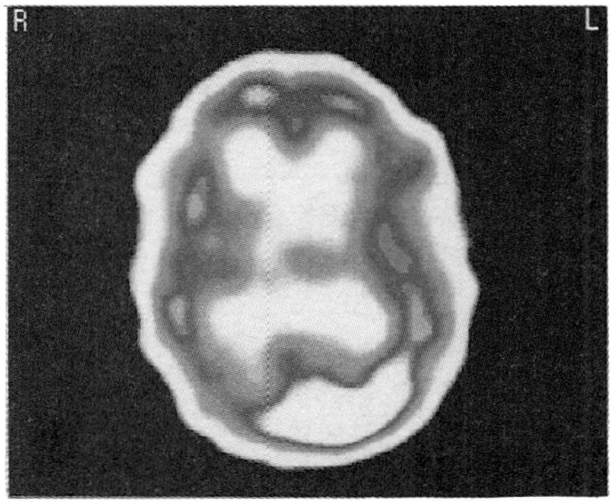

Figure 2.13. HMPAO SPECT. Injection during a complex partial seizure of left occipital origin. This axial slice shows hyperperfusion of the left occipital pole.

1. identifying the abnormalities seen on interictal and perictal scans and the changes seen with respect to the interictal scan;
2. explaining the changes seen by means of a hypothesis of seizure origin and spread.

The first stage primarily requires imaging (nuclear medicine, clinical physics, neuroradiology) expertise, while the second stage requires the input of someone who has experience in epileptology, and in particular understands the pathophysiology of seizure genesis and spread. It is the author's experience that, when changes are complex, it may be necessary to consider more than one possible hypothesis of localization and spread to explain the rCBF changes.

The place of HMPAO SPECT in a programme for the neurosurgery of epilepsy

How should HMPAO SPECT be used in a programme for the surgical treatment of epilepsy? Data in the literature suggest that ictal rCBF changes can be used as a major localizing criterion in mesial temporal seizures, certainly when the treatment offered is temporal lobectomy, rather than a more restricted excision. It would, in the author's opinion, be unnecessary to subject a patient to invasive EEG recordings where ictal HMPAO SPECT had shown hyperperfusion of one temporal lobe, when injection had clearly been carried out during the patient's habitual seizure. One group[34] has already greatly reduced the proportion of their patients undergoing invasive EEG monitoring by using this technique. Mesial temporal lobe seizures represent the majority of adult epilepsies, and this group in itself provides a major indication for use of the technique.

Too few data have been published to allow rCBF SPECT findings as a major localizing criterion for seizures of extratemporal origin, despite promising preliminary results. Moreover, ictal changes may be complex, especially in frontal lobe

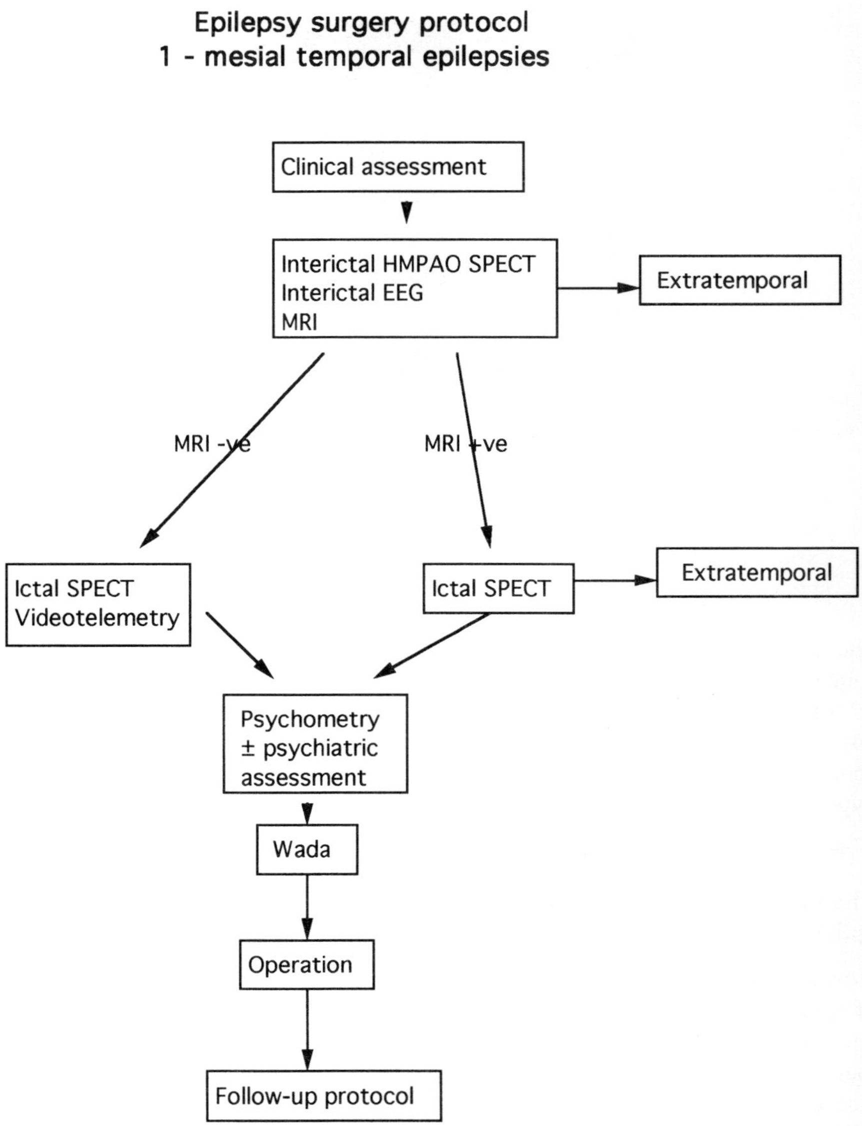

Figure 2.14. The use of HMPAO SPECT in a programme for the neurosurgery of epilepsy. The Glasgow protocol.

seizures, requiring careful interpretation. More needs to be learned of the changes associated with seizures originating in different extratemporal structures. In centres with experience of surgical treatment it is, however, reasonable to add ictal SPECT data to all the other data (clinical, structural imaging, surface ictal and

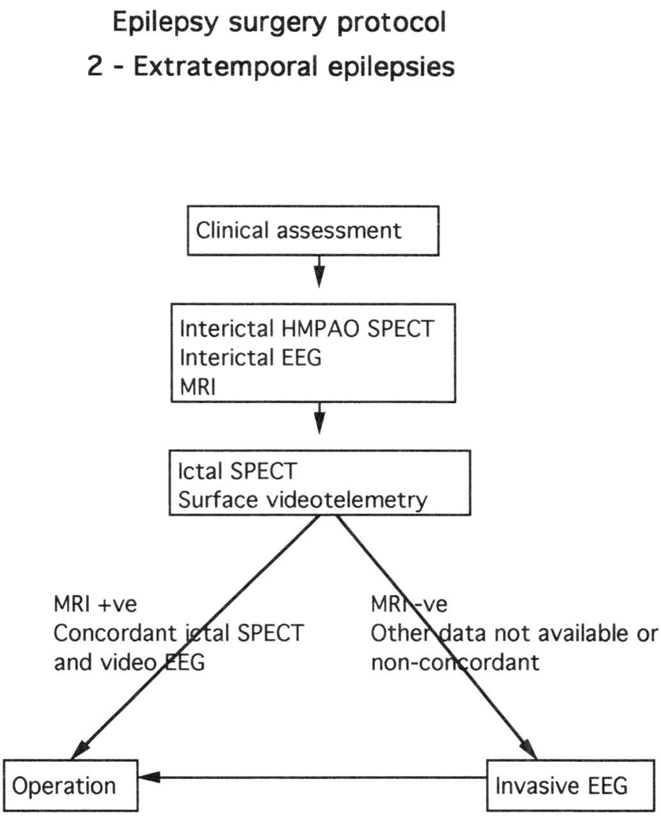

Figure 2.14. *Continued*

interictal EEG) with the hypothesis of seizure origin and spread which allows a logical strategy of invasive EEG monitoring to be constructed.

The current Glasgow presurgical protocol operates on the above principles (Figure 2.14). It places more weight on rCBF SPECT localizing evidence than we would advise in centres with less experience of the technique, particularly if they have relatively little experience of epilepsy surgery and clinical seizure semiology.

SPECT imaging of neurotransmitter receptor function in epilepsy

The imaging of neurotransmitter receptor function is potentially the most exciting area of investigation for SPECT in epilepsy. While HMPAO SPECT can demonstrate the metabolic consequences and location of a seizure, SPECT neurotransmitter receptor studies can potentially demonstrate the abnormalities of neurotransmitter receptor function which underlie abnormal neuronal excitability in epileptogenic zones. Over the past 2–3 decades much evidence has been gathered of the involvement of various neurotransmitter systems in epilepsy, some of which can be investigated using SPECT.

GABA and benzodiazepine receptors

GABA receptors are distributed widely in human brain. The GABA(A) receptor site forms a receptor–ionophore complex which also includes recognition sites for barbiturate, steroid and benzodiazepine (BZ) molecules. Occupation of these sites produces hyperpolarization and neuronal inhibition. Thus, GABA(A) and BZ agonists are anticonvulsant, while antagonists are convulsant. Two subtypes of BZ receptor have been identified. The central type is located on neurones, while the peripheral type is located on glial cells. Studies of slice preparations from animal and human epileptogenic brain tissue have shown alterations in GABA and BZ neurotransmitter function, reductions in GABA terminals and in GABA(A) receptor density[69], reductions in central (neurone associated) BZ receptor density[70].

There is currently no SPECT ligand for the GABA receptor, but the associated BZ site can be imaged with SPECT using [^{123}I]iomazenil. Iomazenil is a partial inverse BZ agonist which binds to central-type BZ receptors. Several small-scale studies have been carried out in epilepsy, showing focal reductions of central type BZ receptor binding in relatively (compared with HMPAO SPECT) high proportions of patients, although there is preliminary evidence that false localizations occur: one study[71] found correct localization in 11 of 13 patients, with one false localization. Preliminary normal data from Glasgow have suggested that low signal from the mesial temporal cortex might exaggerate small differences in BZ binding.

Focal reduction in BZ binding seen on SPECT might be taken to imply a focal change in excitability. Cell loss alone, however, will result in reduced BZ receptor density, possibly without necessarily affecting the balance of excitation and inhibition in the structure concerned. Some early data support the suggestion that BZ binding is merely a sensitive marker for neuronal loss. First, a few false localizing studies have been reported with this technique, as well as some patients in whom there are bilateral reductions of BZ binding but unilateral foci. Both these findings suggest that areas of reduced BZ binding are neither universally found in epileptogenic zones, nor specific to them. Second, reduced BZ binding does not occur in all patients with seizures (no such reductions were found in a series of brain samples from patients with temporal lobe epilepsy secondary to tumours), and so reduced BZ receptor density is presumably not a prerequisite for altered excitability. Third, the limited amount of work performed to relate changes in receptor function to each other and to morphological changes[70] suggest that cell loss alone can account for all or a large proportion of focal loss of BZ binding, in temporal lobe epilepsy at least. If [^{123}I]iomazenil SPECT is merely a very sensitive detector of cell loss, its main competitor in this respect may be MRI, which is at present cheaper, has better anatomical resolution and involves no exposure to ionizing radiation. It is becoming more and more sensitive to subtle lesions and has shown very high sensitivity in some series of patients with temporal lobe epilepsy.

Other neurotransmitter systems

There is considerable evidence that excitatory amino acid neurotransmitters, principally glutamate, are involved in producing the electrical activity and tissue damage associated with seizures[72,73], but there is at present no commercially available SPECT ligand for any element of the excitatory neurotransmitter systems.

Other neurotransmitter systems may be involved in epileptogenesis, in contributory or modulatory ways. Muscarinic agonists can be used to kindle seizures, muscarinic receptor function is abnormal in some animal models of epilepsy[74] and decreased muscarinic receptor binding has been found in human epileptogenic tissue[69]. Stimulation of striatal dopamine D2 receptors inhibits the spread of seizures in animal models[75]. Both these receptors can be imaged using SPECT, using [^{123}I]IBZM and [^{123}I]QNB. Some work is being carried out using another muscarinic agent, [^{123}I]dexetamide, and shows localized abnormalities in temporal lobe epilepsies[76].

SPECT imaging during the intracarotid amytal test

The Wada test is widely used to assess speech and memory function as part of the investigations undertaken prior to temporal lobe surgery. A catheter is placed in the internal carotid artery and contrast is injected to verify the arterial anatomy. Sodium amytal is then injected through the catheter and transiently anaesthetizes those structures of the brain which it perfuses. The test therefore produces a contralateral hemiparesis, and aphasia when the dominant side is injected. The results are used to lateralize speech function and to predict the occurrence of an amnesic syndrome following temporal lobectomy. HMPAO SPECT can be used during the Wada test in two ways. First, structures which are hypoperfused (and therefore are presumably hypofunctioning) as a result of the injection can be imaged; second, one can image the distribution in the brain of the sodium amytal itself.

Interest in this area has arisen from the fact that the Wada test has an intrinsic test–retest variability, and a tendency to produce false-positive results. This has been attributed to variable cross-over of amytal to the contralateral hemisphere via the circle of Willis, which has been shown to occur during contrast injections. However, contrast is not usually injected at the same rate, or in the same volume, as the sodium amytal itself and the distribution of the two has been shown to be different[77]. A clearer picture of the effect of the test can therefore be obtained by imaging the distribution of tissues actually affected by the amytal, or by imaging the distribution of the sodium amytal itself. In the first case, a normal dose of HMPAO is injected intravenously 30–60 s after the onset of the hemiparesis[78],

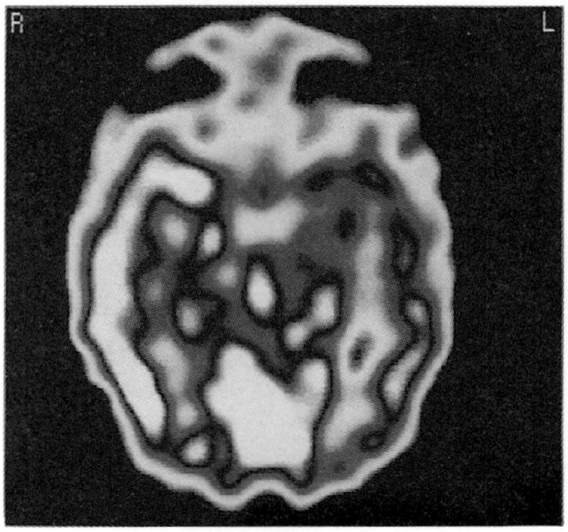

Figure 2.15. HMPAO SPECT carried out during the intracarotid amytal (Wada) test. The patient was injected with 125 mg of sodium amytal into the left carotid artery. Intravenous HMPAO injection was carried out 45 s after the onset of hemiparesis. The image is an axial slice in the long axis of the temporal lobe, and shows hypoperfusion of the left lateral temporal and lateral occipital cortex. In this case, the mesial temporal cortex is hypoperfused.

and acquisition of the image is carried out in the normal way after the amytal test is finished. To trace the distribution on the amytal itself, the HMPAO must be injected intra-arterially through the catheter in the internal carotid artery. Since all the injected dose goes directly to the brain, 5–10% of the usual intravenous dose is sufficient. HMPAO can be injected at the same time as the sodium amytal (i.e. with the two mixed in the same syringe).

Since it is normally perfused by the posterior circulation, the hippocampus (the crucial structure for memory) is not usually perfused with amytal, although variable hypoperfusion of the hippocampus can be seen following intravenous HMPAO injection (Figure 2.15).

HMPAO can be injected intravenously or intra-arterially to verify that appropriate brain structures have been anaesthetized during the Wada test. The data obtained by these methods have suggested flaws in the current physiological rationale for the intracarotid amytal test.

Conclusion

The improving quality of SPECT systems, especially non-head-dedicated systems, means that images with spatial resolution sufficiently good for studies of epilepsy are available at many centres. SPECT data must not, however, be interpreted in isolation from other localizing data, and it is essential that defined image analysis criteria are used. In particular, SPECT images should be compared with an appropriate series of normal scans. Ictal rCBF SPECT is better carried out with concurrent EEG monitoring; if this is not the case images must be interpreted more conservatively. The place of ictal SPECT data in presurgical investigation is gaining wide acceptance for use in individuals with extratemporal as well as mesial temporal seizures. Correlative studies between ictal clinical semiology and ictal rCBF changes are now being undertaken. The ability of rCBF SPECT to see the whole brain provides a perspective distinct from that given by EEG techniques, and promises to improve our understanding of the functional anatomy of the origin and propagation of focal seizures. The more recent development of ligands for neurotransmitter receptors allows in vivo study of neurotransmitter function and, potentially, the effects on this of pharmacological intervention.

Acknowledgement

The SPECT imager used to acquire images in Glasgow patients was funded by the Wellcome Trust. Figures 2.8b, 2.8c and 2.13 were acquired while the author was the recipient of a research fellowship from INSERM, at the unit of Prof. P. Chauvel, INSERM CJF 90-12, Service de Neurologie, CHRU Pontchaillou, Rennes, France. The author also acknowledges the kind and continuing cooperation of the staff of the Nuclear Medicine Department, Centre Eugene Marquis, on the same hospital site.

References

1. Hauser WA, Kurland LT. The epidemiology of epilepsy in Rochester, Minnesota, 1935 through 1967. Epilepsia. 1975;16:1–66.
2. Shorvon SD. Epidemiology, composition, natural history and genetics of epilepsy. Lancet. 1990; 336:93–6.
3. Mazziotta JC, Engel J. The use and impact of positron computed tomography in epilepsy. Epilepsia. 1984;25:S86–104.
4. Devous MD, Leroy RF, Homan RW. Single photon emission computed tomography in epilepsy. Semin Nucl Med. 1990;10:349–56.
5. Horsley V. An address on the origin and seat of epileptic disturbance. Br Med J. 1892;1:693–6.
6. Gibbs FA, Lennox WG, Gibbs EL. Cerebral blood flow preceding and accompanying seizures in man. Arch Neurol Psychiatr. 1934;32:257–72.
7. Penfield W. The circulation of the epileptic brain. Res Publ Ass Nerv Ment Dis. 1937;18: 605–737.
8. Penfield W, von Santha K, Cipriani A. Cerebral blood flow during induced epileptiform seizures in animals and man. J Neurophysiol. 1939;2:257–67.
9. Dymond AM, Crandall PH. Oxygen availability and blood flow in the temporal lobes during spontaneous epileptic seizures in man. Brain Res. 1976;102:191–6.
10. Edvinsson L, MacKenzie ET, McCulloch J. Energy generation in the central nervous system. In: Cerebral blood flow and metabolism. New York: Raven Press; 1990:153–8.
11. Roy CW, Sherrington CS. On the regulation of the blood supply of the brain. J Physiol. 1890; 11:85–108.
12. Kuchinsky W, Wahl M. Local chemical and neurogenic regulation of cerebral vascular resistance. Physiol Rev. 1978;58:656–89.
13. Plum F, Posner JB, Troy B. Cerebral metabolic and circulatory response to induced convulsions in animals. Arch Neurol. 1968;18:1–13.
14. Posner JB, Plum F, Van Poznak A. Cerebral metabolism during electrically induced seizures in man. Arch Neurol. 1969;20:388–95.

15. Tanaka S, Sako K, Tanaka T, Nishihara I, Yonemasu Y. Uncoupling of local blood flow and metabolism in the hippocampal CA3 in kainic acid induced limbic seizures. Neuroscience. 1990;36:339–48.
16. Franck G, Sadzot B, Salmon E, et al. Regional cerebral blood flow and metabolic rates in human focal epilepsy and status epilepticus. In: Delgado-Escueta, et al. editors. Adv Neurology. 1986;44:935–48.
17. Duncan R. Epilepsy, cerebral blood flow and cerebral metabolic rate. Cerebrovasc Brain Metab Rev. 1992;4:105–21.
18. Kuhl DE, Engel J, Phelps ME, Selin C. Epileptic patterns of local cerebral metabolism and perfusion in humans determined by emission computed tomography of 18FDG and 13NH3. Ann Neurol. 1980;8:348–60.
19. Abou Khalil BW, Siegel GJ, Sackellares JC, Gilman S, Hichwa R, Marshall R. Positron emission tomography studies of cerebral glucose metabolism in chronic partial epilepsy. Ann Neurol. 1987;22:480–6.
20. Engel J, Brown WJ, Kuhl DE, Phelps ME, Mazziotta JC, Crandall PH. Pathological findings underlying focal temporal lobe hypometabolism in partial epilepsy. Ann Neurol. 1982;12: 518–28.
21. Theodore WH, Newmark ME, Sato S. [^{18}F]Fluorodeoxyglucose positron emission tomography in refractory complex partial seizures. Ann Neurol. 1983;14:429–37.
22. Sackellares JC, Siegel JG, Abou-Khalil BW, et al. Differences between lateral and mesial temporal metabolism interictally in epilepsy of mesial temporal origin. Neurology. 1990;40: 1420–6.
23. Sperling MR, Gur RC, Alavi A, et al. Subcortical metabolic alterations in partial epilepsy. Epilepsia. 1990;31:145–55.
24. Engel J, Henry TR, Risinger MW, Mazziotta JC, Sutherling WW, Phelps ME. Presurgical evaluation for partial epilepsy: relative contributions of chronic depth electrode recordings versus FDG-PET and scalp-sphenoidal ictal EEG. Neurology. 1990;40:1670–7.
25. Theodore WH, Fishbeinn D, Dubinsky R. Patterns of cerebral glucose metabolism in patients with partial seizures. Neurology. 1988;38:1201–6.
26. Holman BL, Lee RGL, Hill TC, Lovett RD, Lister-James J. A comparison of two cerebral perfusion racers. *N*-Isopropyl I-123 *p*-iodoamphetamine and I-123 HIPDM, in the human. J Nucl Med. 1984;25:25–30.
27. Neirinckx RD, Canning LR, Piper IM, et al. Technetium 99m d,l-HM-PAO: a new radiopharmaceutical for SPECT imaging of regional cerebral blood perfusion. J Nucl Med. 1987;28: 191–202.
28. Rowe CC, Bercovic SF, Sia STB, et al. Localisation of epileptic foci with postictal single photon emission computed tomography. Ann Neurol. 1989;26:660–8.
29. Stefan H, Pawlik G, Bocher-Schwartz HG, et al. Functional and morphological abnormalities in temporal lobe epilepsy: a comparison of interictal and ictal EEG, CT, MRI, SPECT and PET. J Neurol. 1987;234:377–84.
30. Ryding E, Rosen I, Elmqvist D, Ingvar DH. SPECT measurements with 99Tc HMPAO in focal epilepsy. J Cerebr Blood Flow Metab. 1988;8:S95–S100.
31. Bonte FJ, Devous MD, Stokely EM, Homan RW. Single photon tomographic determination of regional cerebral blood flow in epilepsy. Am J Neuroradiol. 1983;4:544–6.
32. Podreka I, Lang W, Suess E, et al. Hexamethyl-propylene-amine-oxime (HMPAO) single photon emission computed tomography (SPECT) in epilepsy. Brain Topography. 1988;1:55–60.
33. Duncan R, Patterson J, Bone I, Wyper DJ, McGeorge AP. Tc99m HMPAO single photon emission computed tomography in temporal lobe epilepsy. Acta Neurol Scand. 1990;81:287–93.
34. Duncan R, Patterson J, Hadley DM, Bone I, Wyper D. SPECT in temporal lobe epilepsy: ictal and interictal studies. Current problems in epilepsy, Vol. 6. London: Libbey; 1990.
35. Duncan R, Patterson J, Hadley DM, et al. CT, MR and SPECT imaging in temporal lobe epilepsy. J Neurol Neurourg Psychiatr. 1990;53:11–5.
36. Rowe CC, Berkovic SF, Austin MC, et al. Visual and quantitative analysis of interictal SPECT with Tc99m HMPAO in temporal lobe epilepsy. J Nucl Med. 1991;32:1688–94.

37. Stefan H, Bauer J, Feistel H, et al. Regional cerebral flow during focal seizures of temporal and fronto central onset. Ann Neurol. 1990;27:162–6.
38. Duncan R, Patterson J, Roberts R, Hadley DM, Bone I. Ictal/postictal SPECT in the pre-surgical localisation of complex partial seizures. J Neurol Neurosurg Psychiatr. 1993;56:141–8.
39. Lang W, Podreka I, Suess E, Muller C, Deecke L. Single photon emission computed tomography during and between seizures. J Neurol. 1988;235:277–84.
40. Sperling B, Lassen NA. Hyperfixation of HMPAO in subacute ischaemic stroke leading to spuriously high estimates of cerebral blood flow by SPECT. Stroke. 1993;24:193–4.
41. Valmier J, Touchon J, Daures P, Zanca M, Baldy Moulinier M. Correlations between cerebral blood flow variations and clinical parameters in temporal lobe epilepsy: an interictal study. J Neurol Neurosurg Psychiatr. 1987;50:1306–11.
42. Duncan S, Gillan J, Duncan R, Brodie M. Interictal HMPAO SPECT: a routine investigation in medically intractable complex partial epilepsy? Epilepsy Res. 1992;13:83–7.
43. Duncan R, Patterson J, Hadley DM, Roberts R, Bone I. Interictal temporal hypoperfusion is related to early onset temporal lobe epilepsy. Epilepsia. 1996;37:134–40.
44. Babb TL, Brown WJ. Neuronal dendritic and vascular profiles of human temporal lobe epilepsy correlated with cellular physiology in vivo. In: Delgado Escueta AV, Ward AA, Woodbury DM, editors. Basic mechanisms of the epilepsies. New York: Raven; 1986.
45. Sagar HJ, Oxbury JM. Hippocampal neuron loss in temporal lobe epilepsy: correlation with early childhood convulsions. Ann Neurol. 1987;22:334–40.
46. Greenamyre JT. The role of glutamate in neurotransmission and in neurologic disease. Arch Neurol. 1986;43:1058–63.
47. Rausch R, Walsh GO. Right hemisphere language dominance in right handed epileptic patients. Arch Neurol. 1984;41:1077–80.
48. Prince DA, Wilder BJ. Control mechanisms in cortical epileptogenic foci, 'surround' inhibition. Arch Neurol. 1967;16:194–202.
49. Engel J, Kuhl DE, Phelps ME, Mazziotta JC. Interictal cerebral glucose metabolism and partial epilepsy and its relation to EEG changes. Ann Neurol. 1982;12:510–7.
50. Biersack HJ, Stefan H, Reichman K. Brain imaging with 99mTc-HMPAO SPECT, CT and NMR – results in epilepsy. J Nucl Med. 1986;27:102.
51. Shen W, Lee BI, Park H, et al. HIPDM-SPECT brain imaging in the presurgical evaluation of patients with intractable seizures. J Nucl Med. 1990;31:1280–4.
52. Harvey AS, Hopkins IJ, Bowe JM, Cook DJ, Shield LK, Berkovic SF. Frontal lobe epilepsy: clinical seizure characteristics and localisation with ictal ^{99m}TcHMPAO SPECT. Neurology. 1993;43:1966–80.
53. Andersen AR, Friberg H, Schmidt JF, Hasselbalch SG. Quantitative measurements of cerebral blood flow using SPECT and Tc99m D,L-HMPAO compared to xenon-133. J Cerebr Blood Flow Metab. 1988;8(Suppl. 1):S69–81.
54. Friberg L, Andersen AR, Lassen NA, Holm S, Dam M. Retention of 99mTc bicisate in the human brain after intracarotid injection. J Cerebr Blood Flow Metab. 1994;14(Suppl. 1):S19–27.
55. Grunwald F, Menzel C, Pavics L, et al. Ictal and interictal brain SPECT imaging in epilepsy using technetium 99mECD. J Nucl Med. 1994;35:1896–901.
56. Newton MR, Bercovik SF, Austin MC, Rowe CC, McKay WJ, Bladin PF. Postictal switch in blood flow distribution and temporal lobe seizures. J Neurol Neurosurg Psychiatr. 1992;55: 891–4.
57. Harvey AS, Bowe JM, Hopkins IJ, Shield LK, Cook DJ, Berkovic SF. Ictal 99mTc HMPAO single photon emission computed tomography in children with temporal lobe epilepsy. Epilepsia. 1993;34:869–77.
58. Lee BI, Markand ON, Wellman HN, et al. HIPDM SPECT in patients with medically intractable complex partial seizures. Arch Neurol. 1988;45:397–402.
59. Berkovic SF, Newton MR, Chiron C, Dulac O. Single photon emission computed tomography. In: Engel JR, Jr., editor. Surgical treatment of the epilepsies. New York: Raven; 1993:233–43.
60. Newton MR, Berkovic SF, Austen MC. Dystonia, clinical lateralisation and regional blood flow changes in temporal lobe seizures. Neurology. 1992;42:371–7.

61. Collins RC, Kennedy C, Sokoloff L, Plum F. Metabolic anatomy of focal motor seizures. Arch Neurol. 1976;33:536–42.
62. Rowe CC, Berkovic SF, Sia STB, Bladin PF. Patterns of postictal blood flow in temporal lobe epilepsy: qualitative and quantitative findings. Neurology. 1991;41:1096–103.
63. Duncan R, Patterson J, Roberts R, Hadley D. Regional cerebral blood flow during posterior seizures: an HMPAO SPECT study. JNNP. 1995;59:203.
64. Marks DA, Katz A, Hoffer P, Spencer SS. Localisation of extratemporal epileptic foci during ictal single photon emission computed tomography. Ann Neurol. 1992;31:250–3.
65. Duncan R, Patterson J, Hadley DM, Roberts R, Bone I. Ictal HMPAO SPECT in frontal lobe seizures. Epilepsia. 1993;34(Suppl. 2):174.
66. Duncan R, Rahi S, Bernard AM, et al. Ictal cerebral blood flow in seizures originating in the posterolateral cortex. J Nucl Med, in press.
67. Ho SS, Berkovic SF, Newton MR, Austin MC, McKay WJ, Bladin PF. Ictal 99mTc HMPAO SPECT findings in parietal lobe epilepsy. Epilepsia. 1993;34(Suppl. 2):112.
68. Newton MR, Berkovic SF, Austin MC, Rowe CC, McKay WJ, Bladin PF. SPECT in the localisation of extratemporal and temporal seizure foci. J Neurol Neurosurg Psychiatr. 1995;59: 26–30.
69. Wyler AR, Nadi NS, Porter RJ. Acetylcholine, GABA, benzodiazepine and glutamate receptors in the temporal lobe of epileptic patients. Neurology. 1987;37(Suppl. 1):S103.
70. Johnson EW, de Lanerolle NC, Kim JH, et al. Central and peripheral benzodiazepine receptors. Neurology. 1992;42:811–5.
71. van Huffelen AC, van Isselt JW, van Veelen CW. Identification of the side of the epileptic focus with ^{123}I Iomazenil SPECT: a comparison with ^{18}FDG and ictal EEG findings in patients with medically intractable complex partial seizures. Acta Neurochir Wien. 1990;50(Suppl. 1):95–9.
72. McDonald JW, Garofalo EA, Hood T, et al. Altered excitatory and inhibitory amino acid receptor binding in hippocampus of patients with temporal lobe epilepsy. Ann Neurol. 1991;29:529–41.
73. Sloviter RS. Excitotoxic mechanisms of epileptic brain damage. In: Schwartz R, Ben-Ari Y, editors. Excitatory amino acids and epilepsy. New York: Plenum; 1985:659–71.
74. Wasterlain CG, Farber DB, Fairchild MD. Synaptic mechanisms in the kindled epileptic focus: a speculative synthesis. Adv Neurol. 1986;44:411–33.
75. al Tajir G, Starr MS. Anticonvulsant effect of striatal dopamine D2 receptor stimulation. Neuroscience. 1991;43:51–7.
76. Boundy KL, Rowe CC, Black AB, et al. Localisation of temporal lobe epileptic foci with I-123 iododexetimide (IDEX) cholinergic neuroreceptor SPECT. Acta Neurol Belg. 1995;95:S39.
77. Jeffery PJ, Monsein LH, Szabo S, et al. Mapping the distribution of amobarbital sodium in the intracarotid Wada test by use of Tc99m HMPAO with SPECT. Radiology. 1991;178:847–50.
78. Ryding E, Sjoholm H, Skeidsvoll H, Elmqvist D. Delayed decrease in hemispheric cerebral blood flow during Wada test demonstrated by 99mTc-HMPAO single photon emission computed tomography. Acta Neurol Scand. 1989;80:248–54.

3. SPECT in head injury

J. T. LINDSAY WILSON and PETER MATHEW

Introduction

It is estimated that, in Britain alone, around 1000 patients per year are discharged from hospital following severe head injury. These survivors are often seriously handicapped, are predominantly young and most will never work again. Nevertheless, their life expectancy is only slightly reduced from normal. In addition to these patients, many others with moderate or mild injuries also have significant morbidity. Exact numbers of patients in these groups are very difficult to estimate, but it is now well established their problems can persist for a considerable length of time. One of the key unresolved issues in head injury is fully characterizing brain damage in survivors.

There is a small but growing literature on the use of SPECT in head injury. Early experience has demonstrated that it may show abnormalities not detected by computerized tomography (CT)[1] or magnetic resonance imaging (MRI)[2]. The field is evolving, and it would be a mistake to attempt a definitive review. Our aim in this chapter is three-fold: to indicate problems associated with blood flow measurement after head injury, to give examples of the kind of novel information that SPECT can provide, and to discuss the possible contribution SPECT may make to understanding and managing head injury.

Limitations of CT and MR

It is something of a paradox that while CT has probably had its greatest clinical influence in the management of head injury, it has substantial deficiencies in characterizing traumatic lesions. CT was introduced in 1973 and became widely available during the 1970s. It proved to be particularly valuable in the detection of haematoma, and its introduction revolutionized the management of this complication of head injury[3]. However, subsequent studies comparing CT and MRI showed that CT is relatively insensitive to many of the lesions present after trauma[4,5]. MRI was introduced in the early 1980s and has since become widely available. MRI is more sensitive than CT to lesions after head injury, particularly non-haemorrhagic lesions[6]. However, it is slower and more expensive than CT, and there are practical problems associated with the magnetic field and the need for patients to be completely enclosed in the imager. MRI is, therefore, more difficult in the acutely ill patient, but is the structural method of choice if the patient is stable and cooperative[7].

R. Duncan (ed): SPECT Imaging of the Brain, 69–93.

Although more sensitive than CT, MRI does not show all abnormalities which are present after head injury. There has thus been interest in imaging modalities which might help to complete the picture of the spectrum of changes present after trauma.

Detection of cerebral blood flow in head injury using ^{133}Xe techniques

Methods of studying regional cerebral blood flow (rCBF) were developed in the early 1960s, long before the advent of CT[8,9]. The intracarotid ^{133}Xe technique is invasive and can usually only be applied to one hemisphere at a time. Methods using the inhalation or intravenous injection of ^{133}Xe are less invasive[10–12] and allow bilateral hemisphere measurements.

The main emphasis of this work is on intracranial changes occurring immediately after injury and on management in the acute stage. Events overtook these CBF techniques, however, with CT becoming established as the method of preference for detecting haematoma and oedema in acute head injury. Nonetheless, the CBF literature remains an invaluable source of information concerning blood flow after trauma. Some key points emerge from this early work which have implications for HMPAO SPECT studies.

Grey matter versus white matter blood flow

The use of ^{133}Xe allowed distinction between blood flow in grey and white matter[11,12]. Enevoldsen and Jensen[13] reported that the identification of fast and slow clearance rates with grey and white matter may break down in acute head injury. They serially studied 23 severe head injuries 3–21 days following injury using the intracarotid ^{133}Xe technique. Clearance curves from oedematous brain had a mono-exponential shape. The interpretation of the latter was that "in severely oedematous tissue, most of the flow in grey matter has approached the flow values of the white matter to such an extent that a distinction between grey and white compartments cannot be made"[13]. In practice, most studies of rCBF in head injury have concentrated on indices of grey matter flow.

Hyperaemia and hypoaemia

A remarkable feature of cerebral blood flow after head injury is that it may show either an overall decrease or an overall increase from baseline levels. Overgaard and Tweed[14] measured CBF in 43 patients using the intracarotid ^{133}Xe method. Poor clinical outcome was observed when patients had either ischaemia (abnormally low flow compared with the group as a whole, $n=5$) or hyperaemia (abnormally high flow, $n=12$) during the acute stage. Obrist et al. confirmed that both reduced and elevated overall levels of blood flow are found early after head injury[15]. They performed serial studies and found that CBF declined in nine patients who subsequently died, while in 25 out of 26 patients who recovered consciousness there was an increase in blood flow. In a subsequent study[16] they found periods of hyperaemia in over half of 75 head-injured patients, and established that in these patients blood flow lost its normal coupling with oxygen metabolism, increasing out of proportion to it. This finding supports the idea that post-traumatic hyperaemia is in excess of metabolic requirements ('luxury perfusion'). Hyperaemia after head injury has also been detected using the nitrous oxide method[17].

To chart the time course of blood flow changes, Obrist[18] studied 14 patients with severe head injury. All patients had a Glasgow Coma Scale score of 7 or less; those with evidence of secondary hypoxic or ischaemic events were excluded. CBF, measured serially at 12 h intervals, was reduced initially and then gradually increased, reaching a hyperaemic peak 57 h after injury. This peak was associated with lowered overall oxygen metabolism and raised intracranial pressure. Blood flow returned to more normal levels at the final examination, 107 h after injury. Thus changes in blood flow after severe head injury appear to follow a typical sequence: flow is low initially, then rises reaching a hyperaemic peak at 2–3 days, decreasing thereafter to near normal levels by 5–6 days following injury.

Blood flow and outcome

Hyperaemia is associated with raised intracranial pressure, and may be a key process in the pathophysiology of acute head injury. However, the significance of hyperaemia for later outcome in survivors is not clear. Although two studies[14,19] found poorer outcome in patients with hyperaemia, most studies have failed to show any relationship between hyperaemia and outcome.

The existence of patients with both abnormally high and abnormally low flows has made it difficult to relate early blood flow to outcome. Jaggi et al. showed that early oxygen metabolism has a more straightforward relationship with outcome: the lower the metabolic rate the poorer the outcome[20]. They argued that blood flow is related to outcome by virtue of its coupling to metabolism, and thus favour oxygen metabolism as a predictor of outcome. However, it should be noted that they found an impressive relationship between CBF and outcome when patients with hyperaemia were excluded. Early low flow was predictive of poorer outcome, and patients with severe disability at follow-up were found to have the lowest blood flow. The study thus implies that, providing hyperaemia and hypoaemia are distinguished, level of overall blood flow may also be a useful predictor of outcome after severe head injury. Late blood flow also appears to correlate with outcome. Barclay et al.[21] measured CBF by the ^{133}Xe inhalation technique in 12 patients from 2 to 13 months after injury and correlated the results with psychometric function. The main conclusion presented was that blood flow deficits persist for at least a year following injury, as do neuropsychological deficits.

Blood flow thresholds

A further important feature of ^{133}Xe rCBF studies is the development of the concept of blood flow thresholds applied to hypoaemia. This concept has been elaborated for head injury by Overgaard et al.[22]. Two threshold levels of cerebral blood flow are proposed: a level necessary for normal function and a level necessary for structural integrity of the tissue. The proposed values of these thresholds are based on an initial slope index of ^{133}Xe washout which in controls gives a mean hemispheric blood flow of 65 ml/100 g/min and a range of regional flows from 40 ml/100 g/min to 99 ml/100 g/min. They suggest that 40 ml/100 g/min represents the lower boundary for normal function, while a value of 20 ml/100 g/min is necessary for structural integrity. In their study they found regional flows below 20 ml/100 g/min in 14 of 23 patients who died or remained in a persistent vegetative state, but in only four of 40 patients who made some form of recovery. Ischaemic flow levels were most common in frontal and parietal regions, corresponding, they argued, to the 'watershed' regions of ischaemic infarction as defined neuropathologically. Presumably such a distribution of ischaemia would relate to low overall flow due to haemodynamic factors and/or raised intracranial pressure, rather than to areas of tissue damage.

CBF and mental state

Acute head injury produces an obvious and dramatic alteration of consciousness, and this may in itself be associated with transient changes in regional CBF. Deutsch and Eisenberg[23] studied 20 patients within 3 days of suffering closed head injury using the ^{133}Xe inhalation technique. Patients showed a frontal hypoperfusion which returned to normal as they recovered consciousness. These authors argue that low frontal CBF may be a short-lasting effect of reduced mental activity after head injury. Changes in rCBF with mental activity are, of course, observed in normal, conscious people. rCBF studies have employed a wide variety of sensory and cognitive challenges[24]. Local blood flow changes of 20–30% during sensory or motor challenge are not uncommon, and even for cognitive activity changes of 10–20% can be demonstrated reliably, showing that mental state is an important variable in determining regional flow. Unfortunately, there has been little work on the effects of cognitive challenge on rCBF in head-injured patients. Wirsén[25] used ^{133}Xe SPECT in a challenge study but found no differences in activation patterns between head-injured patients and controls. However, in this study the changes in CBF reported in both patients and controls were global, suggesting that ^{133}Xe SPECT may be rather insensitive to local changes in blood flow due to mental activity. There is a need for further work on activation patterns in head-injured patients.

^{133}Xe and rCBF: implications for HMPAO SPECT

Studies using ^{133}Xe rCBF techniques imply that five major considerations have a bearing on the interpretation and use of blood flow images obtained using HMPAO SPECT in head injury.

1. *Importance of absolute levels of flow.* CBF work shows the predictive value of absolute levels of overall blood flow, and the significance of global and local blood flow thresholds[20,22]. HMPAO SPECT does not provide absolute levels of flow directly, and this is a limitation of the technique in this area.
2. *Uncoupling of blood flow and metabolism.* In the acute stage blood flow and metabolism can become uncoupled, and areas of metabolic deficit may not be apparent as reduced blood flow[16]. Thus, HMPAO SPECT blood flow images from the acute stage cannot be equated with images of metabolism.
3. *Presence of hyperaemia.* Hyperaemia is relatively common after severe head trauma[14,16,17], and its existence is potentially a confounding factor in SPECT interpretation.
4. *Rapid change in level of blood flow in the acute stage.* Overall blood flow shows quite rapid fluctuations during the first week after severe head injury[18]. The appearance of SPECT blood flow images may consequently be unstable over this period.
5. *Dependence of CBF on mental state.* Some local blood flow changes may be transient and reflect coma rather than an underlying brain abnormality[23]. Blood flow changes are observed with different cognitive tasks in the conscious person[24,26]; such changes are a potential source of variability unless they are properly controlled for.

Given these considerations it is perhaps not surprising that scepticism is sometimes expressed concerning the value of SPECT in head injury. Most of the problems indicated above refer to the acute stage, and it is clear that particular caution should be exercised in interpreting SPECT images acquired in this period. However, CBF techniques also have associated problems. They are limited to measurement of flow values in the cortex, and are prone to errors resulting from the 'look through' phenomenon due to the sensitivity of scintillation crystals on one side of the head to radiation emitted from the contralateral hemisphere[27]. Although, in principle, these techniques measure regional blood flow, the resolution is typically very coarse. Most reports in the literature concern overall blood flow measured unilaterally or bilaterally. Global measures suited the purposes of many CBF studies which primarily addressed the difficult issues surrounding intracranial dynamics in the acute stage.

Stable xenon-enhanced CT CBF

Technology for measuring regional cerebral blood flow using non-radioactive xenon as a contrast agent for CT (Xe/CT CBF) has been clinically available since the early 1980s[28], but the method has only recently been applied to patients with head injury. The technique has some attractive properties, but also some drawbacks. Xe/CT CBF measures absolute levels of blood flow and yields a tomographic image. The images can be directly compared with conventional CT images, and absolute levels of flow can be measured in subcortical structures inaccessible to ^{133}Xe methods[29]. Furthermore, the technique gives accurate measures of low blood flow, making it particularly valuable in defining local brain ischaemia. Drawbacks of the technique include radiation exposure, the anaesthetic properties of xenon, and a low signal-to-noise ratio. High concentrations of xenon produce better signal-to-noise ratios but are potently anaesthetic. At concentrations which are clinically acceptable there is poorer signal-to-noise ratio, and conscious patients experience sensory disturbance.

The focus of Xe/CT CBF studies in head injury is on identifying low blood flow in the acute stage. It is well recognized that ischaemic damage is common in patients with fatal head injuries[30]. However, until recent studies began to concentrate on CBF measurement within the first few hours of injury, very little clinical evidence for the occurrence of ischaemia in severely head-injured patients had been presented. Marion et al.[31] studied 32 severely head-injured patients within 7 days of injury using Xe/CT. They confirmed that global blood flow was often low within the first 24h after injury.

Bouma et al.[32] combined Xe/CT with initial CT scanning and obtained blood flow measurements at a mean of 3h after injury in 35 severely injured patients. They found evidence of ischaemia in 31% of patients: in seven ischaemia was global, and in four it was focal. The occurrence of ischaemia was significantly associated with early mortality, but was not linked to hypoxia or hypertension before resuscitation.

A further, related application of the Xe/CT technique is in the evaluation of the effects of specific intervention in the acute stage. It has been suggested that the use of hyperventilation to control intracranial pressure may sometimes produce regions of cerebral ischaemia. Stringer et al.[33] used the Xe/CT method to study CBF in 12 patients, four of whom had suffered trauma, before and after hyperventilation. Following hyperventilation, apparently normal regions became ischaemic in five patients, regions of hyperaemia became ischaemic in three, and regions of moderate ischaemia became more ischaemic in three. The authors suggest that routine hyperventilation should be avoided, and that it should rather be tailored to the individual patient. An assumption of the use of hyperventilation is that CO_2 vasoresponsivity is preserved. However, a Xe/CT CBF study by Marion and Bouma[34] showed regional variability in CO_2 reactivity after severe head injury. Such variability might account for the unpredictable effects of hyperventilation, and possibly for ischaemia and infarction.

PET studies in head injury

The $H_2{}^{15}O$ PET technique[35] shares with xenon washout the advantage of being able to make repeat measurements, and has the added advantage that it enables three-dimensional tomographic imaging to be undertaken with a resolution of around 10 mm. Unfortunately, it is not widely available, and there have consequently been few studies of PET in head injury.

Rao et al.[36] compared PET with CT in five patients followed up for 3 months to 2 years after head injury and found that PET abnormalities closely corresponded with the site and extent of cerebral dysfunction inferred from neurological and behavioural examinations. CT scans at 2 years showed encephalomalacia and atrophy consistent with earlier PET findings. PET enabled additional parameters such as glucose metabolism to be measured, and showed disturbances in glucose metabolism that extended beyond the structural abnormalities demonstrated by MRI and CT. It also showed disturbances not demonstrated by rCBF measurements using ^{133}Xe[37]. Ruff et al.[38] reported the results of deoxyglucose PET in six head-injured patients scanned 2–5 years after injury, all of whom had had little or no abnormality on CT but substantial neuropsychological deficits. PET corroborated the neuropsychological findings: the most common abnormality was that of reduced frontal flow, possibly a consequence of diffuse injury. Ruff et al.[39] reported similar findings in individuals with mild head injury.

SPECT and CT/MR imaging in trauma

SPECT images, particularly those acquired using high resolution imagers, can be compared directly with the findings of structural imaging using CT or MRI, and the easy availability of SPECT allows comparison of structure and function in large series of patients. Studies relating SPECT to other aspects of head injury are shown in Table 3.1. Comparisons with CT and MRI suggest that a number of patterns of abnormality exist: these are described in detail in the following sections and summarized in Table 3.2. It must be borne in mind that the studies described here represent initial reports of relationships between SPECT and structural imaging.

Early imaging

1. Focal lesion on CT/MR – corresponding SPECT perfusion abnormality:
Early studies using SPECT in patients with acute head injury demonstrated zones of reduced rCBF corresponding to the sites of structural lesions such as haematomas[1,42]. Clearly, the sensitivity of SPECT in detecting small haematomas will depend on the spatial resolution of the imager used. A number of studies have reported that the zone of reduced CBF is larger than the haemorrhagic lesion seen on CT[1,42]. Abdel-Dayem et al.[1] studied 14 patients in coma; CT was carried out within 24 h of head injury and SPECT within 72 h. They found that the pericontusional low density areas seen on CT scans were much smaller than the corresponding SPECT perfusion deficits.

Although SPECT abnormalities may be more extensive than lesions shown on CT, they are not usually larger than focal lesions shown by MRI. A wealth of

Table 3.1. Studies relating SPECT rCBF to other aspects of head injury.

	Patients		SPECT compared with:				
Study	*n*	Time	Method	CT	MR	Outcome	Psychology
Acute							
Abdel-Dayem[1]	14	≤4 d	HMPAO	X		X	
Fumeya[40]	24	≤10 d	IMP	X	X		
Roper[41]	15	≤5 d	HMPAO	X			
Choksey[42]	8	<2 d	HMPAO	X		X	
Bullock[43]	24	≤21 d	Pertech. & HMPAO		X		
Bullock[44]	43	<21 d	HMPAO	X	X		
Sakas[45]	53	≤21 d	HMPAO	X	X	X	
Nedd[46]	16	≤9 d	HMPAO	X			
Post-acute							
Wiedmann[2]	24	5–12 m	HMPAO		X		X
Oder[47]	12	1–4 m and 13–66 m	HMPAO			X	
Maeder[48]	17	1–29 yr	Xe-133		X		
Wirsén[25]	18	1–29 yr	Xe-133			X	X
Gray[49]	53	>6 m	HMPAO	X			
Ichise[50]	29	9 m–15 yr	HMPAO	X	X	X	X
Newton[51]	19	3–36 m	HMPAO	X	X	X	
Oder[52]	36	7–66 m	HMPAO				X
Goldenberg[53]	36	7–66 m	HMPAO				X
Prayer[54]	18	2–36 m	HMPAO	X	X	X	
Acute & Post-acute							
Reid[55]	13	≤2 d and 1–24 m	HMPAO	X		X	
Jacobs[56]	67	<28 d and 3 m	HMPAO	X		X	
Bavetta[57]	10	3–60 d	HMPAO	X	X	X	

d = day, m = month, yr = year, pertech. = pertechnetate

Table 3.2. Patterns of combined CT/MR and SPECT findings in acute head injury.

Focal lesion on CT/MR
- Corresponding SPECT abnormality.
- Adjacent or distant changes on SPECT: focal blood deficits (including diaschisis and contrecoup), hyperaemia, and distant non-focal changes.
- Little or no SPECT abnormality.

CT/MR evidence of diffuse injury
- Abnormal SPECT: e.g. changes in frontal/occipital ratio
- Little or no SPECT abnormality

experimental evidence from animal models and clinical studies demonstrates that contusions are always surrounded by oedema. MRI shows that haemorrhagic contusions seen on CT are associated with a more extensive area of increased enhancement on T2 weighted images[4,5]. Bullock et al.[44] found that the SPECT defect corresponds to the area of T2 enhancement on MR which is associated with contusions seen on CT, but that the region of low blood flow is no larger than this area of enhancement. The SPECT defect clearly represents haematoma together with the effect of peri-lesional oedema. Serial CTs show that approximately one-third of contusions are associated with an increasing mass effect from oedema[58]. The extent of rCBF defect corresponding to a structural lesion may therefore be related to the timing of the scanning.

2. Focal lesion on CT/MR – adjacent or distant SPECT changes:
a. Adjacent or distant blood flow deficits: Abdel-Dayem et al.[1] reported that 23 focal lesions seen on SPECT were not shown by CT, out of a total of 70 focal lesions detected with both imaging modalities. In a study by Roper et al.[41] corresponding figures were 17 lesions out of a total of 44 lesions in 15 patients scanned acutely. Lesions seen only on SPECT were evaluated with respect to the site of initial impact of the injury; seven of 17 (41%) of the lesions were said to represent 'contrecoup' injuries. In both studies, the number of lesions was derived by counting the number seen on each slice, but there are problems in interpreting figures for number of lesions derived in this way. A given abnormality may appear on more than one slice, and it can be difficult to determine whether they represent distinct lesions. On the same slice, lesions may be irregular and separation can be difficult.

It is possible that many of the distant rCBF focal deficits not seen on CT have an MRI correlate, but data on comparison with MRI early after injury is limited. Fumeya et al.[40] performed imaging within 10 days of injury and reported that of 13 lesions seen on MRI but not on CT, three showed normal and 10 showed reduced flow. The assumption is that these lesions represent areas of focal oedema, seen on MRI but not CT. Recent work in Glasgow comparing acute MR and SPECT carried out within 5 days of head injury has shown distant focal flow deficits in areas with no corresponding MR abnormality in 10 of 30 patients. Two of the abnormalities could be regarded as 'contrecoup' injuries[59]; an example is shown in Figure 3.1 (see also colour section).

b. Focal hyperaemia: Focal hyperaemia has been reported in acute head injury by several authors[40,44,59]. Hyperaemia appears to be associated with focal contusions, but the precise spatial relationship to haemorrhagic areas shown on CT and MRI is a matter of debate. Fumeya et al.[40] suggested that hyperaemic lesions may appear as areas of high intensity in T2-weighted MRI. In their study, hyperaemia was evident only in lesions seen on both CT and MRI. They suggest that SPECT may allow the distinction between oedema (with low flow) and contusion (with hyperaemia). On the other hand, Bullock[44] found that hyperaemia does not occur in areas which are abnormal on CT or MR, but only affects

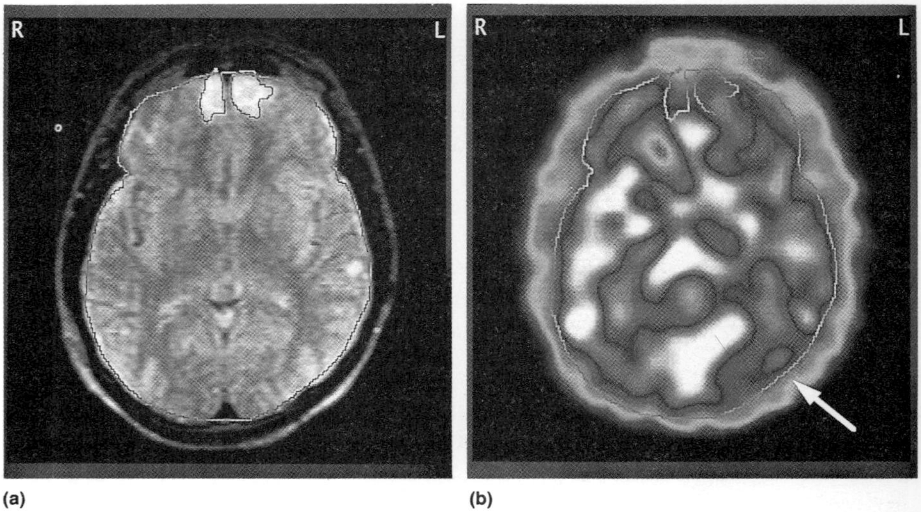

Figure 3.1 Images from a 23-year-old woman who fell from a loft. She was conscious on day 5, but very uncooperative and aggressive. (a) T2 weighted MRI shows bifrontal lesions (outlined). (b) On the SPECT image there are matching frontal flow deficits; there is also a 'contrecoup' left occipital perfusion defect with no matching structural lesion (arrow). (For colour version of this figure see page 183.)

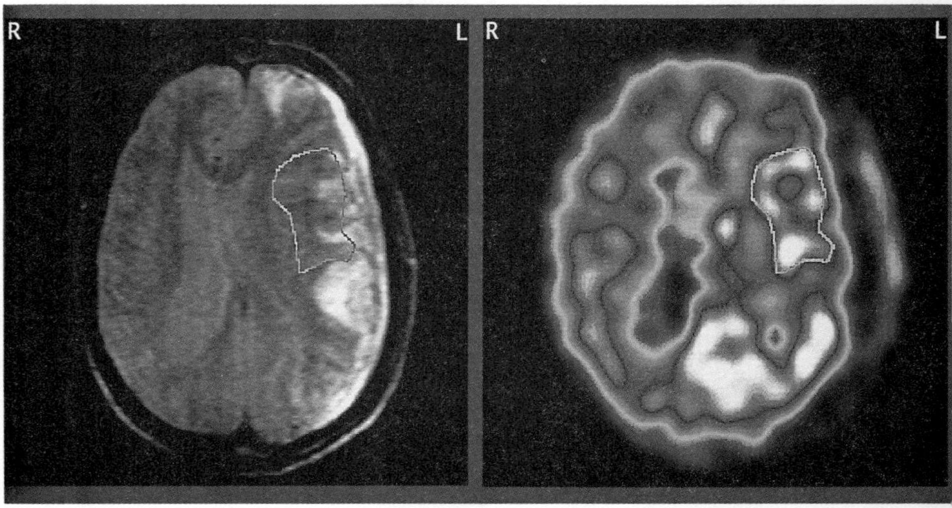

Figure 3.2 A patient with head injury of moderate severity complicated by a left subdural haematoma. MRI and HMPAO SPECT were performed 5 days following injury. The MRI image shows the subdural collection, with an underlying contusion, seen as an area of increased T2 signal just inferior to the marked region of interest. The corresponding SPECT image shows the physical effects of the haematoma and a small area of hypoperfusion in association with the contusion. There is, however, marked hyperperfusion anterior to the contusion which falls within MRI-normal tissue (as marked by the ROI placed on both co-registered images), and a similar area posterior to the contusion. (For colour version of this figure see page 183.)

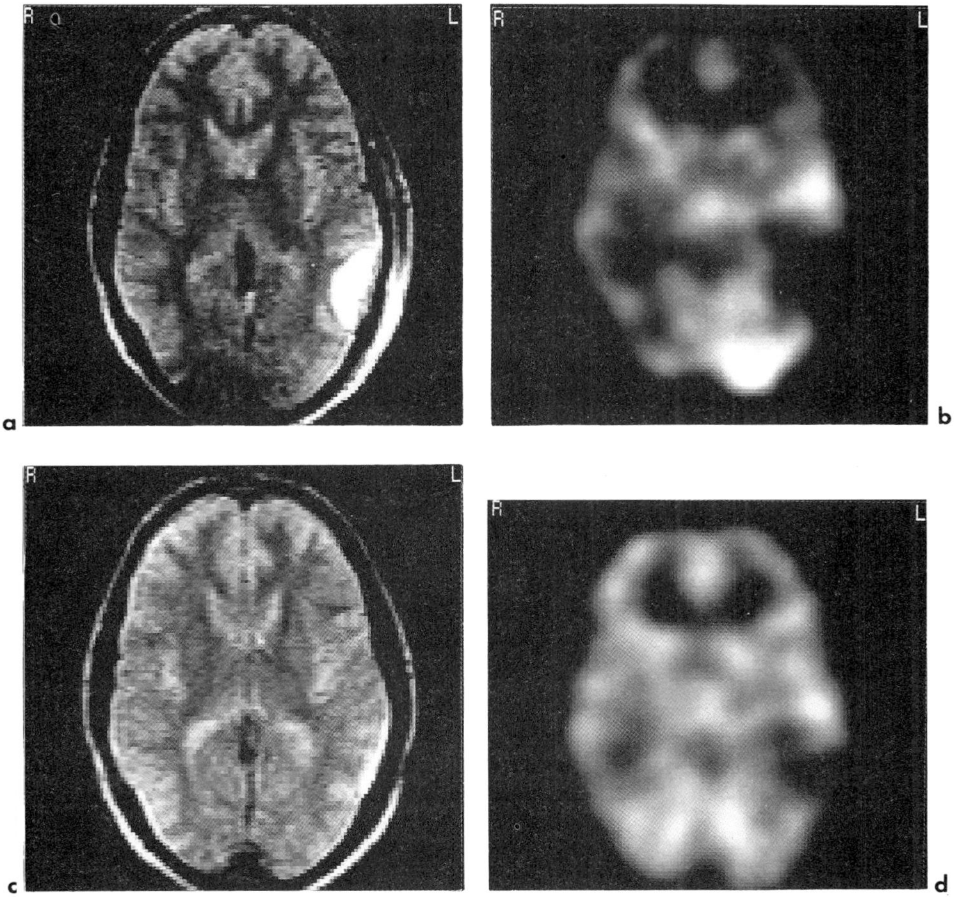

Figure 3.3 A 44-year-old man who had been assaulted. He was conscious on admission, but was acutely aphasic. By 6 months after injury he had made a good recovery. (a) T2 weighted MR images 2 days after injury show an extensive left temporal lesion. (b) SPECT images at the same time show a region of low blood flow corresponding to the lesion shown on MRI. There are adjacent areas of hyperaemia in the left occipital and temporal regions. Blood flow in the frontal regions also appears relatively low. (c) T2 weighted MR images at 6 months show a much smaller abnormality. (d) On SPECT at 6 months there is still a small corresponding perfusion defect, however, the distant abnormalities have disappeared. There is good agreement between MRI and SPECT in the size and location of the contusional abnormality. Acute SPECT shows adjacent and distant changes in blood flow, not seen on MRI.

surrounding tissue. Figure 3.2 (see also colour section) illustrates a case where MR normal tissue adjacent to subdural haematoma and contusion is hyperperfused. Zones of focal hyperaemia were demonstrated in 42% of patients scanned within 2 weeks of injury. These zones were adjacent to SPECT perfusion deficits, in tissue that appeared normal on CT and MR. A region of hyperfixation of HMPAO is also shown in Figure 3.3 which corresponds to the description given by Bullock[44], and has resolved on the late scan. Sakas et al.[45] have extended the follow up in this study, finding that hyperaemia is not associated with poor outcome.

c. Non-focal blood flow abnormalities: Distant non-focal changes associated with trauma may reflect diffuse injury or hemisphere damage. A review of papers in which distant non-focal SPECT findings are reported indicates the following three categories:

I. Asymmetrical right/left hemispheric perfusion[1,42,46]. Choksey et al.[42] reported that in seven of eight patients with focal lesions on CT there appeared to be low perfusion in the rest of the CT-defined structurally normal ipsilateral hemisphere. Nedd and colleagues[46] found asymmetrical perfusion in seven of eight patients who had sustained skull fractures and who showed no parenchymal lesions on CT.
II. Change in the fronto/occipital ratio[1,41]. Frontal perfusion is normally approximately the same as flow in the occipital regions. A relative decrease in frontal flow has been reported by Abdel-Dayem[1] and by Roper[41]. Low frontal flow in association with a focal lesion is evident in Figure 3.3.
III. Diffusely decreased cerebral perfusion[1,47]. As already noted, absolute CBF techniques show that head injury often results in an overall reduction in flow. Despite the technical limitations of SPECT in this regard, some workers have reported a diffuse reduction in perfusion. Objective reporting criteria for diffuse reduction in flow have yet to be defined.

3. Focal lesions on CT/MR – no abnormality on SPECT:
In the study by Roper and colleagues[41] CT demonstrated 12 lesions not seen on SPECT, out of a total of 44 lesions. Of these two were subarachnoid haemorrhage, two were thin acute subdural haematomas but eight were contusions. Abdel-Dayem et al.[1] report that SPECT missed meningeal lesions, defined as sub-arachnoid haemorrhage or subdural haematomas, and one focal lesion, out of a total of 70 lesions. Clearly, acute subarachnoid haemorrhage (uncomplicated by vasospasm and ischaemia) and thin subdural haematomas exerting little mass effect do not produce a perfusion deficit measurable within the spatial resolution of current SPECT scanners. One explanation offered by Roper[41] for the failure of SPECT to pick up larger intraparenchymal lesions may be the dissociation between CBF and reduced metabolic demand in these lesions, that is a relative hyperaemia, which makes them 'invisible' on SPECT. Serial scanning was not carried out in this study; it is likely that the amount of dissociation will vary in time and this may be responsible for the discrepancy in lesion number between SPECT and CT in patients scanned with differing injuries at different time points. An example of a patient who had little or no abnormality on acute SPECT, but a substantial lesion on MRI is shown in Figure 3.4

4. Diffuse injury on CT/MR – SPECT abnormalities:
Roper and colleagues[41] found diffuse swelling as defined by CT in 12 of 15

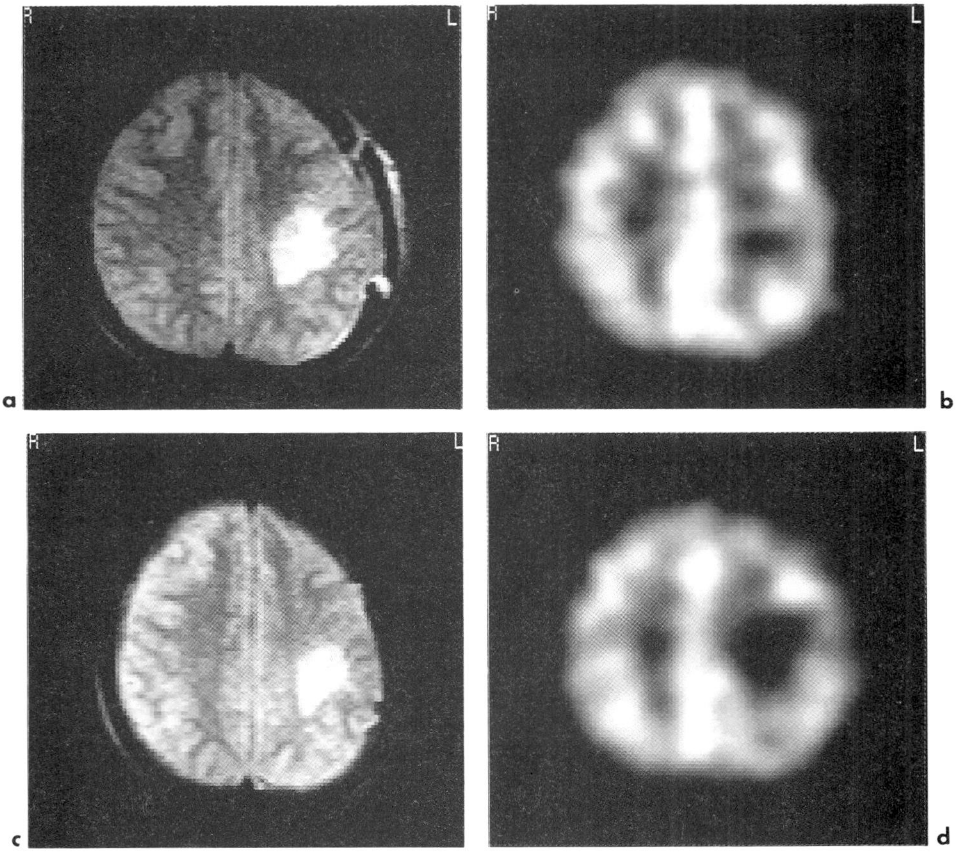

Figure 3.4. Images from a 24-year-old man who suffered a sports injury. He underwent surgery for haematoma on the day of injury. At 6 months he had a marked right sided hemiparesis, but no cognitive deficits. (a) T2 weighted MRI 9 days post-injury shows a large left fronto-parietal lesion. (b) SPECT imaging at 9 days shows little abnormality. (c) T2 weighted MRI at 6 months shows a lesion similar in size to that seen on the acute scan. (d) On SPECT there is a corresponding perfusion defect. SPECT in this case was apparently normal initially even though there was a clear lesion on MRI. The initial SPECT finding implies that in the acute stage there is still significant blood flow in the damaged region: such preserved blood flow may represent a possible window for therapeutic intervention.

patients. In corresponding HMPAO SPECT scans, change in the fronto-occipital slope was present in four patients, but no correlation was found between the degree of swelling on CT and slope reversal.

There are no systematic studies comparing acute CT and rCBF in cases of hypoxic damage, or severe diffuse injury. However, a case is illustrated in Figure 3.5. Acute MRI in this patient showed lesions at grey–white matter interfaces, consistent with diffuse axonal injury. Early HMPAO SPECT shows an obvious reduction in frontal flow.

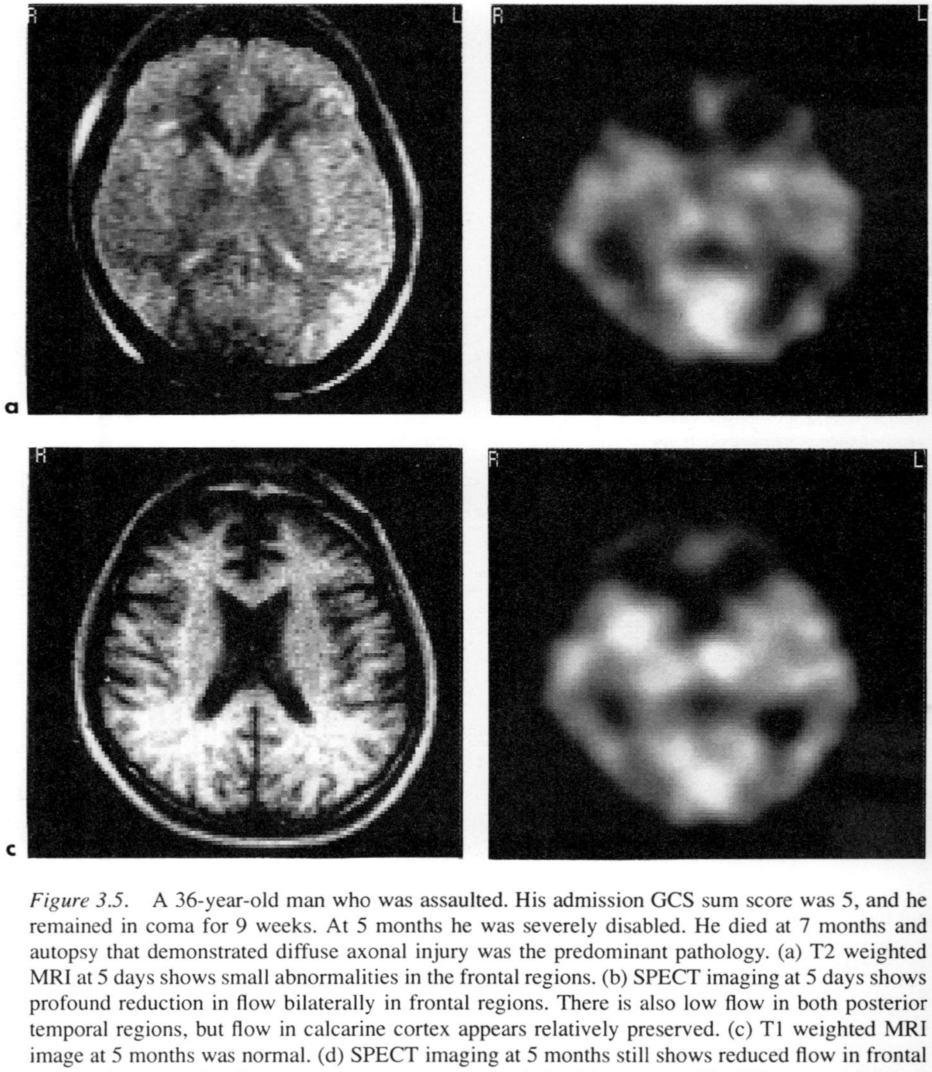

Figure 3.5. A 36-year-old man who was assaulted. His admission GCS sum score was 5, and he remained in coma for 9 weeks. At 5 months he was severely disabled. He died at 7 months and autopsy that demonstrated diffuse axonal injury was the predominant pathology. (a) T2 weighted MRI at 5 days shows small abnormalities in the frontal regions. (b) SPECT imaging at 5 days shows profound reduction in flow bilaterally in frontal regions. There is also low flow in both posterior temporal regions, but flow in calcarine cortex appears relatively preserved. (c) T1 weighted MRI image at 5 months was normal. (d) SPECT imaging at 5 months still shows reduced flow in frontal regions, but posterior flow now appears more normal. In this patient with diffuse axonal injury SPECT reveals a much more extensive abnormality than shown by MRI. SPECT imaging agreed with the poor clinical state of this patient.

Late imaging

1. Focal lesions on CT/MRI:

Gray et al.[49] scanned 53 patients more than 6 months after head injury and demonstrated 42 defects (focal and/or diffuse) by SPECT compared with only 29 focal deficits on CT. In a comparative study between CT, MRI and SPECT Newton et al.[51] showed that SPECT demonstrated more focal lesions than either

CT or MRI, alone or in combination. Wiedmann et al.[2] reported that eight patients with focal head injury scanned more than 6 months after trauma showed differences in location of abnormalities on SPECT and MRI: all patients showed abnormalities on SPECT in the orbito-frontal and frontal areas while MRI detected abnormalities in more patients in anterior temporal and parietal areas. Late imaging of focal lesions is illustrated in Figures 3.3 and 3.4.

2. Diffuse injury on CT/MRI:
Wiedmann et al.[2] found that 11 of 12 patients with MRI-defined diffuse injury had SPECT abnormalities, but as in the focal group (see above) there was considerable variation between the scans in the actual location of abnormalities. SPECT identified abnormalities in the parietal lobes in more patients, whereas MRI detected abnormalities in the orbito-frontal and anterior temporal regions in more patients. Late imaging of a diffuse lesion is shown in Figure 3.5. Prayer et al.[54] followed up 18 severely head-injured patients who had normal early CT. In nine patients who had MR evidence of diffuse axonal injury on follow up, HMPAO SPECT found low flow in cortical and subcortical areas, most often in frontal, temporal and parietal lobes, and in the thalamus.

Sensitivity of HMPAO SPECT to traumatic lesions

HMPAO SPECT provides information which is different from that given by CT and MRI in both acute and follow-up studies of head injury. In specific circumstances SPECT reveals more abnormalities than either CT or MR, but the general claim that SPECT is more sensitive should be treated with caution. In addition to the types of injury involved, differences in the spatial resolution of the imaging modalities being compared may contribute to discrepancies between modalities: CT or MRI may well show small abnormalities which are simply below the resolution of SPECT, even when low-field MRI is being compared with high resolution SPECT.

Variation in reporting practice may also be problematic. For SPECT, unlike CT or MRI, there are no generally agreed criteria for the definition of an abnormality, which can be somewhat arbitrary. In many studies abnormalities are identified by visual inspection, either by comparing images with a control group, or by comparing equivalent areas on right and left sides in the same individual. Large blood flow abnormalities can be unequivocally identified by this procedure, but smaller variations in blood flow may in some centres be interpreted as abnormal, but in others be regarded as being within normal limits. One study with a normal control group found that variations in symmetry of blood flow of up to 18% are not uncommon[60], yet such variations would usually be regarded as 'clinically significant'. Thus adopting too lax criteria for abnormality on SPECT may lead to an unacceptable rate of false positives. For these reasons then, general comparisons between modalities should be made with caution. Studies which precisely align and measure abnormalities seen on different imaging modalities are just beginning to be carried out.

SPECT does, nonetheless, seem to be more sensitive than CT early after head injury, with more lesions demonstrated, particularly in those with minor head injury. The few studies in which MRI and SPECT have been compared in the early phase suggest that SPECT does not detect more focal lesions, the lesion number and size seen on the latter correlating well with not only the haemorrhagic contusion but also the oedema that surrounds the lesions. Considering focal and diffuse SPECT changes together, SPECT may show more abnormalities than MRI in the early phase. This is illustrated by Figure 3.3: on MRI there is a single large abnormality in the acute stage, and a rather smaller abnormality at later follow-up. On the other hand, acute SPECT shows a range of abnormalities, including low flow in the region of the MRI lesion, and adjacent hyperaemia. At follow up, there is only low flow on SPECT.

Studies later after head injury also show differences. The study by Gray et al.[49] suggest that SPECT scanning more than 6 months after the injury can demonstrate regional brain dysfunction when CT results are normal. Wiedmann et al.[2] found that 50% of blood flow abnormalities identified by SPECT were independent of the sites of structural damage identified by MRI, showing that HMPAO SPECT can provide information not offered by current structural imaging techniques.

Applications of HMPAO SPECT in head injury

A variety of clinical applications have been proposed for SPECT imaging (Table 3.3). At present there is too little information to evaluate such proposals fully, but some candidates appear to emerge as stronger than others. There seems to be a place for use of HMPAO SPECT in establishing brain death[55]. However, this need not involve the use of SPECT imaging as such: HMPAO can be used with radionuclide angiography for this purpose[64]. In the acute stage HMPAO SPECT has proved particularly useful in evaluating focal lesions, and achieving a better understanding of concomitant CBF changes. In post-acute studies SPECT appears to hold the promise of showing correlates of behavioural and psychological disturbance.

SPECT and the pathophysiology of acute closed head injury

Contusions are regarded by many as the hallmark of brain injury, occurring in approximately 90% of cases[65]. As cortical impact lesions, usually occurring in non-eloquent areas of brain (eg. sub-frontal and temporal pole regions) and therefore not associated with immediate focal neurological deficit, they assume clinical importance when pericontusional oedema progresses and causes an increase in intracranial pressure with consequent deterioration in conscious level[65,66]. Operation to evacuate a mass lesion is then required. Acute subdural haematoma, the most common mass lesion which complicates severe head

Table 3.3. Possible clinical applications of SPECT imaging after traumatic head injury.

Evaluation of focal lesions[41]
Predicting late deterioration in haematoma[42]
Evaluation of diffuse lesions[47,48]
Predicting outcome[56,61]
Predicting degree of permanent brain damage[55]
Diagnosis of brain death[55]
Predicting recovery from the persistent vegetative state[47]
Evaluating treatment effects on the vegetative state[62]
Evaluation of pharmaceutical intervention
Predicting susceptibility to post-traumatic epilepsy[63]
Predicting post-traumatic headache[55]
Evaluation of mild head injury[46,49]
Assessment of outcome[51]
Revealing correlates of behaviour disorders[52]

injury, is often associated with massive hemisphere swelling: 60% of such patients die or remain severely disabled[30].

The pathophysiological mechanisms that initiate and sustain the acute perilesional swelling seen in focal intradural traumatic lesions are largely unknown. In addition, the effects of these potentially treatable secondary events on long-term neuropsychological and behavioural outcome remain to be determined.

Only a few studies have reported serial SPECT scans in the early phase. These demonstrate that the SPECT perfusion deficit may not necessarily increase with time, and indeed, may reduce in size[44,67]. Choksey et al.[42] suggest that the degree of mismatch between haematoma size seen on CT and the deficit on SPECT (ie. the amount of pericontusional oedema around the haematoma) may predict late deterioration, but numbers in the study were small. This may be merely a reflection of the size of the haemorrhagic component of the contusion; Statham and Todd[68] have shown that the amount of oedema is related to haematoma volume.

Bullock et al.[67] mapped blood–brain barrier changes using ^{99m}Tc pertechnetate in 13 patients with focal changes on CT/MRI within 30 days of injury. Perilesional blood–brain barrier disturbance was present in all patients with contusion ($n=7$), intracranial haematoma ($n=2$), and acute subdural haematoma ($n=4$), although changes were most marked in contusions. In five of seven of these patients undergoing HMPAO SPECT scanning at a similar time, a corresponding zone of hypoperfusion was present. A further study demonstrated that blood–brain barrier breakdown around contusions was more frequent after the first 48 h after injury[43]. A similar temporal pattern was seen with acute subdural haematoma. In a study by Hadley et al.[68], using MRI and gadolinium, pericontusional enhancement on T1 weighted images occurring in a zone of hypointensity on a non-contrast scan (due to leakage of paramagnetic contrast agent from the intravascular compartment to the surrounding brain parenchyma in areas of increased blood–brain barrier permeability) was only seen at least 4 days following injury.

It is clear that while mechanical disruption of the blood–brain barrier occurs at the time of injury (and can presumably be demonstrated by hyperacute gadolinium MRI or pertechnetate SPECT) other mechanisms operate to produce delayed blood–brain barrier damage, leading to the concepts of early cytotoxic perilesional oedema, and later or delayed vasogenic oedema[69]. The underlying mechanisms of persisting or progressive hypoperfusion on SPECT therefore change with time. In the region of perilesional oedema, compression of vessels by the haematoma causes a reduction in local tissue perfusion, possibly exacerbated by perivascular astrocyte swelling and formation of endothelial microvilli. Metabolic requirements of damaged neurones may also be reduced. Many of these microscopic and ultrastructural features resolve within a few days and cellular swelling is replaced by an increase in the extracellular compartment. Putative mechanisms for the delayed increase in permeability of the blood–brain barrier (and possible exacerbation of neuronal damage) include the generation of free radicals, either as a consequence of reperfusion in previously nearly ischaemic vessels as resolution/healing of the lesion takes place or as part of an acute inflammatory reaction[70,71].

With the improved spatial resolution of recent scanners, and new software that allows relatively straightforward and accurate registration/comparison between MRI and SPECT it is now possible to measure relative blood flow in MRI-defined regions of interest (Figure 3.1). Controlled clinical trials in Glasgow are in progress to measure the effect of intervention with specific neuroprotective agents (eg. NMDA antagonists and free radical antagonists), using endpoints of blood flow in the lesion proper, in the perilesional region, and in the ipsilateral and contralateral distant hemisphere; the role of specific mechanisms may be clarified by such studies.

SPECT shows discrete areas of hyperperfusion adjacent to perfusion deficits, and distant non-focal changes in patients with less severe head injury. It therefore seems that a focal structural lesion occurring as part of a moderate or minor head injury may still be associated with widespread distant functional changes. These are clearly distinct from the centrifugal sequence of structural changes which occur in severe acceleration–deceleration injuries, in which rotational forces produce damage to deep midline regions and brain stem[72]. Experiments with animal models have suggested that focal cortical injury may be associated with widespread histopathological and ultrastructural hemispheric changes in the absence of evidence of diffuse axonal injury[73]. Propagation of shear forces from the point of impact and/or diffusion of 'toxic substances' from the damaged tissue may cause functional changes with no structural counterpart. Determination of whether these functional changes persist and contribute to eventual outcome requires studies involving acute and follow-up imaging. Preliminary studies in Glasgow have shown that the more minor perfusion deficits seen on early scanning, and lesions which appear on SPECT only, are more likely to resolve[74]. Recovery of function with time may suggest minor metabolic derangement as opposed to the more severe diffuse axonal injury in which neurones are deafferented and fail to recover.

SPECT and psychological outcome in chronic head injury

Newton et al.[51] and Bavetta et al.[57] have demonstrated that relatively simple measures derived from early and late SPECT imaging can show good agreement with outcome. In the study by Newton et al. Glasgow Outcome Scale category showed correlations both with number of SPECT lesions ($r = 0.82$) and global blood flow ($r = 0.74$): the most severely disabled patients showed the highest number of lesions and the lowest cerebral blood flow. One implication of the study is that overall blood flow and apparent number of abnormalities are related.

Wiedmann et al.[2] included detailed neuropsychological testing in their study. As already noted, patients with focal mass lesions had, proportionally, a greater number of abnormal regions of interest than patients with diffuse injuries. However, those with focal lesions were less severely injured by GCS or PTA criteria, and showed less evidence of neuropsychological impairment than the diffuse group. This finding underscores the importance of distinguishing different subtypes of head injury when relating imaging findings to outcome. A small abnormality after a diffuse injury may be associated with poorer outcome than an extensive abnormality due to a focal mass[75]. Wiedmann et al.[2] investigated relationships between blood flow deficits and neuropsychological test performance in each subgroup separately. There were few significant associations in the focal group, perhaps because this group showed little neuropsychological impairment. However, in the diffuse group nine of the 12 neuropsychological measures were related to blood flow abnormalities.

The relationship between cognitive and psychosocial problems after trauma and HMPAO SPECT was investigated in studies by Goldenberg et al.[53] and Oder et al.[52]. The latter studied 36 very severely injured survivors: the median duration of post-traumatic amnesia was 2 months. SPECT was performed using a dual head rotating gamma camera, and analysis was conducted by computing counts for regions of interest normalized by the mean for the whole brain. Goldenberg et al.[53] administered a neuropsychological battery which emphasized memory and executive functions. The authors did not find the expected relationships between memory functions and temporal blood flow, and between executive functions and frontal blood flow. There were general relationships between test performance and blood flow in frontal and thalamic areas, with the latter providing the strongest correlations ($r \approx 0.5$).

These findings are similar to those reported for MRI studies attempting to localize specific psychological deficits to temporal and frontal regions[76]. It would seem that for SPECT, as for MRI and CT, establishing differential psychological effects of frontal and temporal abnormalities is problematic. In general, the strongest relationships between psychological deficits and CT or MRI are found for measures which appear to reflect diffuse damage[77]. Ichise et al.[50] compared two measures indicative of overall brain damage; the anteroposterior blood flow ratio (APR) derived from SPECT, and the ventricle–brain ratio (VBR) derived from MRI. Relationships with neuropsychological test performance were stronger for the APR than the VBR, particularly in patients with minor head injury.

Oder et al.[52] investigated behavioural and psychosocial problems in the same patients as Goldenberg et al.[53]. Factor analysis of personality testing indicated that social isolation, disinhibition and aggressive behaviour were the main post-traumatic changes in patients. The highest correlation ($r=0.6$) was found between frontal lobe blood flow and disinhibition: low flow was associated with greater disinhibition. Social isolation was associated with low flow in left hemisphere regions, and aggressive behaviour with low right hemisphere flow, although these correlations were weak ($r \approx 0.4$). There was also a weak correlation between neurological and cognitive deficits and thalamic flow. These results lend support to the idea that disinhibition syndrome (disinhibited behaviour, childishness and euphoria) after head injury is due to orbito-frontal damage[78].

Taken together, the studies by Goldenberg et al.[53] and Oder et al.[52] suggest that personality change has a stronger relationship with blood flow than does cognitive deficit. Low frontal flow appears to be specifically related to personality change and psychosocial problems. The results also suggest that blood flow in frontal and thalamic regions may be indicative of the degree of diffuse damage.

SPECT in minor head injury

In recent years there has been a great deal of interest in the idea that there may be persisting psychological problems after apparently minor head injury. In such cases there is often a discrepancy between CT findings, which are normal, and neuropsychological assessment, which shows impairment. It would thus be of great significance to find an imaging test which reliably demonstrated an organic substrate for the psychological complaints of these patients.

Gray et al.[49] studied 53 patients who had been injured 6 months or more previously. Analysis of images was by visual inspection in comparison to a group of controls. rCBF abnormalities were found in 42 patients (80%), while only 29 (55%) had abnormal CT. SPECT was particularly successful in demonstrating abnormalities in patients with minor head injury: 60% had SPECT abnormalities, compared with 25% on CT (in this study minor head injury was defined as a loss of consciousness of 20 min or less, a GCS of 13–15, and hospital stay for 2 days or less). Nedd and colleagues[46] report results from 16 patients with mild to moderate injuries. CT and HMPAO SPECT imaging were carried out within the first 9 days of injury; GCS scores ranged from 11 to 15. A SPECT abnormality was defined as a variation in symmetry of 20% or more on two contiguous slices. SPECT was abnormal in 14 (88%) patients while CT was abnormal in six (38%).

These studies are interesting, but they have obvious weaknesses. As noted already, false positive results are a problem for SPECT studies, particularly when variations in blood flow are small. Nedd and colleagues[46] candidly point out that, in the absence of structural abnormality, there is often not "sufficient data to characterize a 'decrease in rCBF' definitely as a 'lesion'". The studies would gain more weight if it could be shown that SPECT abnormalities were related to psychological deficits, but neither study included psychological measures.

Two recent studies shed further light on minor head injury in this respect. Jacobs et al.[56] studied a group of 67 patients, including 25 with mild head injury. All patients who had normal rCBF early after injury were asymptomatic at 3 months, while six of the nine patients who had abnormal rCBF early after injury had clinical signs or symptoms. The authors conclude that normal rCBF in the early phase is a predictor of favourable outcome, and that SPECT can provide an objective correlate of the complaints made by patients with minor head injury. Ruff et al.[39] carried out a PET study of nine patients who had persisting neuropsychological problems after minor head injury. They were selected as having little or no abnormality on structural imaging. PET showed abnormalities of rCBF in all the patients, most commonly in the frontal lobes.

It seems, then, that persisting neuropsychological deficits following minor head injury do correlate with abnormal rCBF. Whether such abnormalities are secondary to subtle structural damage to which current structural imaging methods are insensitive, are a form of 'functional brain damage'[50], or whether they represent changes in mental state arising from other causes, such as depression, is currently unknown. Caution should therefore currently be exercised in the interpretation of SPECT images in minor head injury when findings are unsupported by structural imaging.

Conclusion

Early CBF and more recent rCBF SPECT studies have contributed significantly to understanding the pathophysiology of head injury, and to the long term link between cerebral blood flow and brain function. The main emerging clinical application is to provide objective correlates for behavioural and psychometric abnormalities found following head injury in individuals who have normal results on structural imaging.

Acknowledgement

Work carried out in Glasgow on SPECT and MR in head injury is supported by the Wellcome Trust.

References

1. Abdel-Dayem HM, Sadek SA, Kouris K, et al. Changes in cerebral perfusion after acute head injury: comparison of CT with Tc-99m HM-PAO SPECT. Radiology. 1987;165:221–6.
2. Wiedmann KD, Wilson JTL, Wyper D, Hadley DM, Teasdale GM, Brooks DN. SPECT cerebral blood flow, MR imaging, and neuropsychological findings in traumatic head injury. Neuropsychology. 1989;3:267–81.

3. French BN, Dublin AB. The value of computerized tomography in the management of 1000 consecutive head injuries. Surg Neurol. 1977;7:171–83.
4. Hadley DM, Teasdale GM, Jenkins A, et al. Magnetic resonance imaging in acute head injury. Clin Radiol. 1988;39:131–9.
5. Snow RB, Zimmerman RD, Gandy SE, Deck MDF. Comparison of magnetic resonance imaging and computed tomography in the evaluation of head injury. Neurosurgery. 1986;18: 45–52.
6. Gentry LR, Godersky JC, Thompson B. MR imaging of head trauma: Review of the distribution and radiopathologic features of traumatic lesions. Am J Neuroradiol. 1988;9: 101–10.
7. Zimmerman RA, Bilaniuk LT. CT and MR: Diagnosis and evolution of head injury, stroke, and brain tumors. Neuropsychology. 1989;3:191–230.
8. Lassen NA, Ingvar DH. Blood flow of cerebral cortex determined by radioactive krypton-85. Experientia. 196;17:42–5.
9. Harper AM, Glass HI, Steven JL, Granat AH. The measurement of local blood flow in the cerebral cortex from the clearance of xenon 133. J Neurol Neurosurg Psychiatry. 1964;27: 255–8.
10. Mallett BL, Veall N. Investigation of cerebral blood flow in hypertension using radioactive xenon inhalation and extracranial recording. Lancet. 1963;i:1081–2.
11. Obrist WD, Thompson HK, King CH, Wang HS. Determination of regional cerebral blood flow by inhalation of 133-xenon. Circ Res. 1967;20:124–35.
12. Obrist WD, Thompson HK, Wang HS, Wilkinson WE. Regional cerebral blood flow estimated by ^{133}xenon inhalation. Stroke. 1975;6:245–56.
13. Enevoldsen EM, Jensen FT. Compartmental analysis of regional cerebral blood flow in patients with acute severe head injuries. J Neurosurg. 1977;47:699–712.
14. Overgaard J, Tweed WA. Cerebral circulation after head injury. Part 1: cerebral blood flow and its regulation after closed head injury with emphasis on clinical correlations. J Neurosurg. 1974;41:531–41.
15. Obrist WD, Gennarelli TA, Segawa H, Dolinskas CA, Langfitt TW. Relation of cerebral blood flow to neurological status and outcome in head injury. J Neurosurg. 1979;51:292–300.
16. Obrist WD, Langfitt TW, Jaggi JL, Cruz J, Gennarelli TA. Cerebral blood flow and metabolism in comatose patients with acute head injury. J Neurosurg. 1984;61:241–53.
17. Robertson CS, Contant CF, Gokaslan ZL, Narayan RK, Grossman RG. Cerebral blood flow, arteriovenous oxygen difference, and outcome in head injured patients. J Neurol Neurosurg Psychiatry. 1992;55:594–603.
18. Obrist WD, Marion DW, Aggarwal S, Darby JM. Time course of cerebral blood flow and metabolism in comatose patients with acute head injury. J Cerebr Blood Flow Metab. 1993; 13(Suppl. 1):A571.
19. Uzzell BP, Obrist WD, Dolinskas CA, Langfitt TW. Relationship of acute CBF and ICP findings to neuropsychological outcome in severe head injury. J Neurosurg. 1986;65:630–5.
20. Jaggi JL, Obrist WD, Gennarelli TA, Langfitt TW. Relationship of early cerebral blood flow and metabolism to outcome in acute head injury. J Neurosurg. 1990;72:176–82.
21. Barclay L, Zemcov A, Reichert W, Blass JP. Cerebral blood flow decrements in chronic head injury syndrome. Biol Psychiatry. 1985;20:146–57.
22. Overgaard J, Mosdal C, Tweed WA. Cerebral circulation after head injury. Part 3: does reduced regional cerebral blood flow determine recovery of brain function after blunt head injury? J Neurosurg. 1981;55:63–74.
23. Deutsch G, Eisenberg HM. Frontal blood flow changes in recovery from coma. J Cerebr Blood Flow Metab. 1987;7:29–34.
24. Friberg L, Roland PE. Functional activation and inhibition of regional cerebral blood flow and metabolism. In: Olesen J, Edvinsson L, editors. Basic mechanisms of headache. Amsterdam: Elsevier, 1988:89–98.
25. Wirsén A. Chronic effects of traumatic frontal lesions: Clinical, neurophysiological and neuropsychological aspects [Doctoral dissertation]. University of Lund, Sweden, 1991.

26. George MS, Ring HA, Costa DC, Ell PJ, Kouris K, Jarritt PH. Neuroactivation and neuroimaging with SPET. London: Springer, 1991.
27. Wyper DJ, Cooke MBD. Application of lateral counting corrections when measuring cerebral blood flow. Phys Med Biol. 1978;23:1130–40.
28. Johnson DW, Stringer WA, Marks MP, Yonas H, Good WF, Gur D. Stable xenon CT cerebral blood flow imaging: Rationale for and role in clinical decision making. Am J Neuroradiol. 1991;12:201–13.
29. Harrington TR, Manwaring K, Hodak J. Local basal ganglia and brain stem blood flow in the head injured patient using stable xenon enhanced CT scanning. In: Miller JD, Teasdale GM, Rowan JO, Galbraith SL, Mendelow AD, editors. Intracranial pressure VI. Berlin: Springer, 1986:680–6.
30. Graham DI, Ford I, Adams JH, et al. Ischaemic brain damage is still common in fatal nonmissile head injury. J Neurol Neurosurg Psychiatry. 1989;52:346–50.
31. Marion DW, Darby J, Yonas H. Acute regional cerebral blood flow changes caused by severe head injuries. J Neurosurg. 1991;74:407–14.
32. Bouma GJ, Muizelaar PJ, Stringer WA, Choi SC, Fatouros P, Young HF. Ultra-early posttraumatic regional CBF measurement. J Neurosurg. 1992;77:360–8.
33. Stringer WA, Hasso AN, Thompson JR, Hinshaw DB, Jordan KG. Hyperventilation-induced cerebral ischemia in patients with acute brain lesions: demonstration by xenon-enhanced CT. Am J Neuroradiol. 1993;14:475–84.
34. Marion DW, Bouma GJ. The use of stable xenon-enhanced computed tomographic studies of cerebral blood flow to define changes in cerebral carbon dioxide vasoresponsivity caused by severe head injury. Neurosurgery. 1991;29:869–73.
35. Lammertsma AA, Wise RJS, Cox TCS, Thomas DGT, Jones T. Measurement of blood flow, oxygen utilisation, oxygen extraction ratio, and fractional blood volume in human brain tumours and surrounding oedematous tissue. Br J Radiol. 1985;58:725–34.
36. Rao N, Turski PA, Polcyn RE, Nickels RJ, Mathews CG, Flynn MM. ^{18}F Positron emission computed tomography in closed head injury. Arch Phys Med Rehab. 1984;65:780–5.
37. Langfitt TW, Obrist WD, Alavi A, et al. Computerized tomography, magnetic resonance imaging, and positron emission tomography in the study of brain trauma: preliminary observations. J Neurosurg. 1986;64:760–7.
38. Ruff RM, Buchsbaum MS, Tröster AI, et al. Computerized tomography, neuropsychology, and positron emission tomography in the evaluation of head injury. Neuropsych Neuropsychol Behav Neurol. 1989;2:103–23.
39. Ruff RM, Crouch JA, Troster AI. Selected cases of poor outcome following minor brain trauma: comparing neuropsychological and positron emission tomography assessment. Brain Injury. 1994;8:297–308.
40. Fumeya H, Ito K, Yamagiwa O, et al. Analysis of MRI and SPECT in patients with acute head injury. Acta Neurochir (Wien). 1990;(Suppl. 51):283–5.
41. Roper SN, Mena I, King WA, et al. An analysis of cerebral blood flow in acute closed head injury using technetium-99m-HMPAO-SPECT and computed tomography. J Nucl Med. 1991; 32:1684–7.
42. Choksey MS, Costa DC, Iannotti F, Ell PJ, Crockard HA. ^{99}Tc^{m}-HMPAO SPECT studies in traumatic intracerebral haematoma. J Neurol Neurosurg Psychiatry. 1991;54:6–11.
43. Bullock R, Statham P, Patterson J, Wyper D, Hadley D, Teasdale E. The time course of vasogenic oedema after focal human head injury – evidence from SPECT mapping of blood brain barrier defects. Acta Neurochir (Wien). 1990;(Suppl. 51):286–8.
44. Bullock R, Sakas D, Patterson J, et al. Early post-traumatic cerebral blood flow mapping: correlation with structural damage after focal injury. Acta Neurochir (Wien). 1992;(Suppl. 55):14–7.
45. Sakas DE, Bullock MR, Patterson J, Hadley D, Wyper DJ, Teasdale GM. Focal cerebral hyperaemia after focal head injury in humans: a benign phenomenon? J Neurosurg. 1995;83: 277–84.
46. Nedd K, Sfakiankis G, Ganz W, et al. ^{99m}Tc-HMPAO SPECT of the brain in mild to moderate

traumatic brain injury patients: compared with CT – a prospective study. Brain Injury. 1993;7: 469–79.
47. Oder W, Goldenberg G, Podreka I, Deecke L. HM-PAO-SPECT in persistent vegetative state after head injury: prognostic indicator of the likelihood of recovery? Intensive Care Med. 1991;17:149–53.
48. Maeder P, Wirsén A, Bajc M, et al. Volumes of chronic traumatic frontal brain lesions measured by MR imaging and CBF tomography. Acta Radiol. 1991;32:271–8.
49. Gray BG, Ichise M, Dae-Gyun C, Kirsh JC, Franks W. Technetium-99m-HMPAO SPECT in the evaluation of patients with a remote history of traumatic brain injury: a comparison with X-ray computed tomography. J Nucl Med. 1992;33:52–8.
50. Ichise M, Chung DG, Wang P, Worzman G, Gray BG, Franks W. Technetium 99 HMPAO SPECT in the evaluation of patients with chronic traumatic brain injury – a correlation with neuropsychological performance. J Nucl Med. 1994;35:217–26.
51. Newton MR, Greenwood RJ, Britton KE, et al. A study comparing SPECT with CT and MRI after closed head injury. J Neurol Neurosurg Psychiatry. 1992;55:92–4.
52. Oder W, Goldenberg G, Spatt J, Podreka I, Binder H, Deecke L. Behavioural and psychosocial sequelae of severe closed head injury and regional cerebral blood flow: A SPECT study. J Neurol Neurosurg Psychiatry. 1992;55:475–70.
53. Goldenberg G, Oder W, Spatt J, Podreka I. Cerebral correlates of disturbed executive function and memory in survivors of severe closed head injury: A SPECT study. J Neurol Neurosurg Psychiatry. 1992;55:362–8.
54. Prayer L, Wimberger D, Oder W. Cranial MR imaging and cerebral 99mTc HMPAO SPECT in patients with subacute or chronic severe closed head injury and normal CT examinations. Acta Radiol. 1993;34:593–9.
55. Reid RH, Gulenchyn KY, Ballinger JR, Ventureyra ECG. Cerebral perfusion imaging with technetium-99m HMPAO following cerebral trauma: Initial experience. Clin Nucl Med. 1990; 15:383–8.
56. Jacobs A, Put E, Ingels M, Bossuyt A. Prospective evaluation of technetium 99m HMPAO SPECT in mild and moderate traumatic brain injury. J Nucl Med. 1994;35:942–7.
57. Bavetta S, Nimmon CC, White J. A prospective study comparing SPET with MRI and CT as prognostic indicators following severe closed head injury. Nucl Med Commun. 1994;15: 961–8.
58. Kobayashi S, Nakazawa S, Otsuk T. Clinical value of serial computer tomography with severe head injury. Surg Neurol. 1983;20:25–9.
59. Mathew P, Hadley DM, Condon B, Teasdale GM. Blood brain barrier permeability in head injured patients – effects of tirilazad mesylate. J Neurotrauma. 1993;10(Suppl. 1):S164.
60. Hodgson M, Smith DJ, Macleod MA, Houston AS, Francis TJR. Case control study of cerebral perfusion deficits in divers using $^{99}Tc^{m}$ hexamethylpropylene amine oxime. Undersea Biomed Res. 1991;18:421–31.
61. Tikovsky RS. Predicting outcome in traumatic brain injury: what role for rCBF spect? J Nucl Med. 1994;35:947–8.
62. Kanno T, Kamei Y, Yokoyama T. Treating the vegetative state with dorsal column stimulation. In: Takakura K, Kanno T, editors. The Society for Treatment of Coma, Volume 1. Tokyo: Neuron Publishing, 1993:67–76.
63. Dalmady-Israel C, Zasler ND. Post-traumatic seizures: a critical review. Brain Injury. 1993;7: 263–73.
64. Laurin NR, Driedger AA, Hurwitz GA, et al. Cerebral perfusion imaging with technetium-99m HM-PAO in brain death and severe central nervous system injury. J Nucl Med. 1989;30: 1627–35.
65. Adams JH, Gennarelli TA, Graham DI. Brain damage in non-missile head injury: Observations in man and subhuman primates. Recent Adv Neuropathol. 1980;2:165–90.
66. Miller JD, Becker DP, Ward DJ, Sullivan HG, Adams WE, Rosner MJ. Significance of intracranial hypertension in severe head injury. J Neurosurg. 1977;47:510–6.
67. Bullock R, Statham P, Patterson J, Teasdale GM, Teasdale E, Wyper D. Tomographic map-

ping of CBF, CBV and blood brain barrier changes in humans after focal head injury using SPECT – mechanisms for late deterioration. In: Hoff JT, Betz AL, editors. Intracranial pressure VII. Berlin: Springer Verlag, 1989.
68. Statham PFX, Todd NV. Intracerebral haematoma: aetiology and haematoma volume determine the amount and progression of brain oedema. Acta Neurochir (Wien). 1990;(Suppl. 51):289–91.
68. Hadley DM, Macpherson P, Lang DA, Teasdale GM. Breaches in the blood brain barrier in head injured patients shown by gadolinium enhanced magnetic resonance imaging. Neuroradiology. 1991;(Suppl. 33):86–8.
69. Bullock R, Maxwell WL, Graham DI, Teasdale GM, Adams JH. Glial swelling following human cerebral contusion: an ultrastructural study. J Neurol Neurosurg Psychiatry. 1991;54: 427–34.
70. Chan PH, Long NS, Fishman RA. Protective effects of liposome-entrapped superoxide dismutase on post-traumatic brain oedema. Ann Neurol. 1987;21:540–7.
71. Kochanek PM, Hallenbeck JM. Polymorphonuclear leukocytes and monocytes/macrophages in the pathogenesis of cerebral ischaemia and stroke. Stroke. 1992;23:1367–79.
72. Ommaya AK, Gennarelli TA. Cerebral concussion and traumatic unconsciousness. Correlation of experimental and clinical observations on blunt head injuries. Brain. 1975;97:633–54.
73. Maxwell WL, Follows R, Ashurst DE, Berry M. The response of the cerebral hemisphere of the rat to injury I. The mature rat. Phil Trans R Soc Lond. 1990;B328:479–500.
74. Mitchener A, Wyper DJ, Mathew P, et al. SPECT imaging in head injury: the relationship between acute abnormalities and residual clinical and cognitive deficits at 6 months. J Cerebr Blood Flow Metab. 1993;13(Suppl. 1):S545.
75. Wilson JTL, Wyper D. Neuroimaging and neuropsychological functioning following closed head injury: CT, MR, and SPECT. J Head Trauma Rehabil. 1992;7:29–39.
76. Levin HS, Williams DH, Eisenberg HM, High WM Jr, Guinto FC Jr. Serial MRI and neurobehavioural findings after mild to moderate closed head injury. J Neurol Neurosurg Psychiatry. 1992;55:255–62.
77. Wilson JTL. The relationship between neuropsychological function and brain damage detected by neuroimaging after closed head injury. Brain Injury. 1990;4:349–63.
78. Stuss DT, Benson FD. The frontal lobes. New York: Raven Press, 1986.

4. SPECT in cerebrovascular disorders

DONALD G. GROSSET and IAN BONE

Introduction

Stroke represents a major public health problem with an incidence of 100–300 per 100 000 of the population a year. In the USA there are 3 million stroke survivors and 500 000 new first strokes per annum. Stroke accounts for 10% of all deaths in Western countries, age-adjusted mortality in the USA being 50–100 per 100 000 of the population. Ischaemic stroke is four times more common than haemorrhagic stroke, accounting for 70–80% of all cases. Intracranial haemorrhage accounts for 20–30%, of which one-third to one-half are the consequence of subarachnoid haemorrhage. Stroke risk is determined by age, ethnic origin, genetic predisposition and modifiable risk factors such as hypertension, associated cardiac disease and cigarette smoking. Mortality varies from 8–20% in the first 30 days in ischaemic stroke, depending on specific stroke subtype. The annual recurrence rate for first ever stroke varies between 6 and 9%. Ischaemic cerebrovascular disease is therefore a common disorder, presenting not at specialist centres but to local community practitioners and district hospitals. It is important that potential investigative techniques are available to the majority of sufferers: SPECT systems capable of producing images of sufficient quality to be useful in the context of cerebrovascular disease are now available even in relatively small centres.

Since the innovative work of Kety and Schmidt in the late 1940s there has been a continuing and developing interest in the physiology of cerebral blood flow (CBF) in both health and disease. Logically, cerebrovascular diseases have been widely studied. Quantitative techniques applied to the study of cerebral haemodynamics initially involved inhalation and later intracarotid injection of radioactive inert gases. Techniques involving intracarotid injection of radiolabelled albumin or microsphere angiography have been developed but have failed to find widespread application in clinical practice. Reivich et al.[1] first demonstrated the quantitative evaluation of regional cerebral blood flow and cerebral metabolism of glucose and oxygen using PET. The lack of accuracy and reproducibility of early techniques and the complexity, invasiveness and cost of later ones have precluded their widespread use in the clinical setting, but the development of SPECT has allowed the application of functional imaging to cerebrovascular disease in routine clinical practice. SPECT ligands for neurotransmitter receptors have not been fully evaluated in cerebrovascular disease, and only

R. Duncan (ed): SPECT Imaging of the Brain, 95–130.

rCBF tracers have been widely used. The two tracers most widely used are iodine isopropyl 123 iodoamphetamine (IMP) and technetium-99m hexamethyl propylene amine oxime ([^{99m}Tc]HMPAO). A further tracer ^{99m}Tc ethyl cystein dimer (ECD) is currently being evaluated[2,3]. IMP is relatively unstable in brain, and so requires early scanning to generate meaningful images of rCBF. With this tracer depression of plasma pH, as occurs in brain ischaemia, may result in underestimations of regional blood flow changes. Despite these drawbacks the tracer was the first applied in clinical studies of cerebrovascular disease and is still widely used. [^{99m}Tc]HMPAO can be labelled on site, allowing emergency unscheduled studies to be performed. It is currently available and approved for clinical studies in most countries and in view of its longer post-injection stability is the tracer of preference in studies of rCBF in cerebrovascular disease.

How might estimation of regional cerebral blood flow in suspected ischaemic stroke aid diagnosis and management? In ischaemic stroke rCBF SPECT will almost immediately show a perfusion defect that correlates with eventual computerized tomography (CT) and magnetic resonance imaging (MRI) changes. It may, therefore, allow early detection and definition of cerebral infarction, confirming diagnosis at a stage when initiation of thrombolytic or cytoprotective treatment might beneficially influence outcome. SPECT could be used to monitor such treatment and could be used in conjunction with transcranial Doppler to demonstrate and quantify the presence and degree of reperfusion, with prognostic implications. SPECT combined with cranial CT or MRI may contribute to a clearer delineation of clinical syndromes associated with specific vascular territory occlusions and may be of value in explaining the remote effects of infarction, for example the presence of cortical features such as neglect or language dysfunction in patients with CT or MRI subcortical infarcts. Studies of regional cerebral blood flow under stress such as CO_2 inhalation or following intravenous acetazolamide can yield 'dynamic' information on cerebrovascular reactivity and the collateral circulation (Figures 4.1, 4.2). This could contribute to the confirmation and understanding of the haemodyamic significance of stenotic or occlusive extracranial disease. Such stress tests may be used to monitor and select patients for surgical reconstructive or revascularization procedures. One major challenge in subarachnoid haemorrhage is the early detection of vasospasm. SPECT in conjunction with transcranial Doppler may allow early identification of this problem and the selection of patients at risk for specific trial therapy. SPECT may also aid in the diagnosis of vascular dementia. Finally, the development of ligands for neuroreceptors such as 5-hydroxytryptamine (5HT) and glutamate may enhance our knowledge of biochemical changes in cerebral ischaemia and even monitor the cytoprotective effect of these receptor antagonists in clinical trials.

We will go on to examine the work that is the basis of these possible applications of SPECT in cerebrovascular diseases.

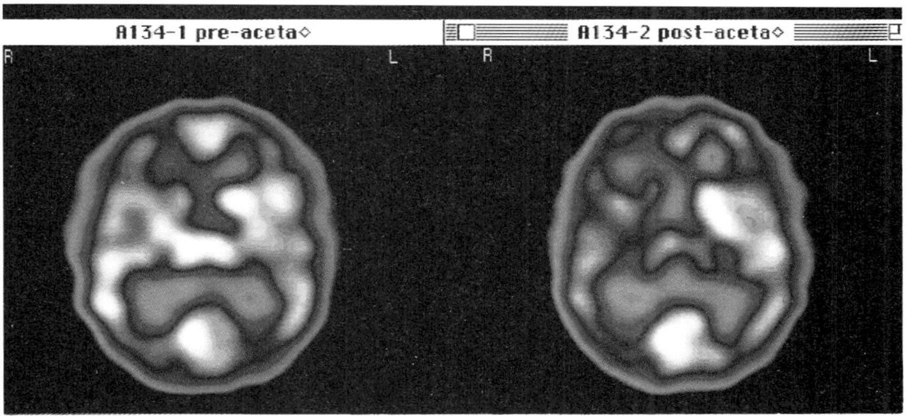

Figure 4.1. SPECT HMPAO scans in a 45-year-old woman with an aortic arch syndrome with right carotid occlusion. The initial scan taken before intravenous injection of 1 g acetazolamide shows a symmetrical pattern for cerebral blood flow. The second scan taken after intravenous acetazolamide shows an area of hyporeactivity throughout the anterior right hemisphere in territories supplied by the terminal branches of the right internal carotid artery.

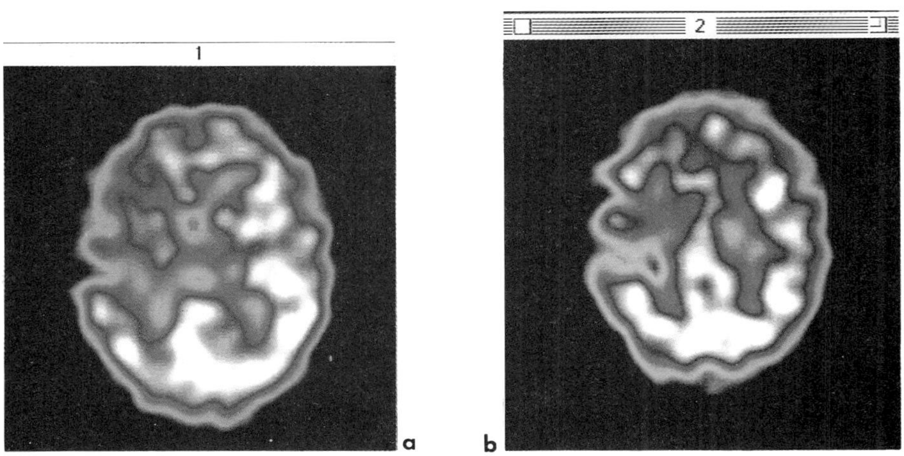

Figure 4.2. A 55-year-old man who sustained a right internal carotid artery occlusion confirmed by carotid angiography. The initial scan (a) shows cerebral blood flow prior to CO_2 inhalation; it can be seen that blood flow is significantly diminished throughout the right hemisphere. Following 5% CO_2 inhalation (b) there is a significant increase in regional cerebral blood flow throughout the whole of the left hemisphere with an increase on the right side also, but to a lesser extent. This demonstrates that whereas cerebral vascular reactivity is retained in both hemispheres it is clearly impaired on the side of the carotid occlusion.

Transient ischaemic attack (TIA)

A TIA is a reversible episode of focal neurological dysfunction due to transient failure of brain perfusion. The incidence is 30 per 100 000 of the population and approximately one-third of patients suffering a TIA will experience a disabling or fatal stroke within 5 years. Approximately 10% of all stroke sufferers can recount an antecedent TIA, and 25% of patients have a cardiac embolic source. A larger proportion have disease of the extracranial vessels, ranging from atheromatous plaque to severe stenosis or occlusion of the internal carotid artery or vessels of the posterior circulation. The likelihood of stroke after TIA is increased by the presence of occlusive or severe stenotic extracranial vascular disease, and the small proportion of patients who have no demonstrable cause appear to have a good prognosis. TIAs in the posterior circulation have a similar mechanism but are generally regarded as being more benign, though this has recently been questioned. TIAs are occasionally the consequence of haemodynamic rather than embolic mechanisms: intermittent cardiac dysrhythmias (causing reduced cardiac output) associated with local occlusive or stenotic extracranial vascular disease may present as an episode of focal neurological dysfunction. Once a diagnosis of TIA is made, investigation should be directed towards identification of the cause, distinguishing embolic from flow related mechanisms.

Early studies of rCBF in patients with TIAs, such as that of Skinhoj et al.[4], employed inhalation or intracarotid injection of radiolabelled inert gases. Some such studies found that the initial disturbance of blood flow was restored to normal within hours. Others claimed a persisting reduction lasting some weeks. These early studies used different methodologies, applied them to small numbers of patients and were not combined with structural imaging to detect the eventual development of structural abnormality. They also failed to take account of the presence or relevance of occlusive or stenotic extracranial vascular disease. SPECT studies in hemisphere TIAs have in general shown a 60% reduction of tracer uptake on day one, 40% reduction on day two and then a gradual return to normal within a period of one week. Such findings perhaps are only to be expected in a condition by which focal perfusion failure produces transient symptoms that rapidly recover with restoration of perfusion and without the development of infarction. For example, Bogousslavsky et al.[5] studied 12 patients with a >75% stenosis of the internal carotid artery and a single ipsilateral hemisphere transient ischaemic event of less than 1 h duration. All patients had normal cranial CT, and HMPAO SPECT was performed 0–50 h after the event. Eight patients showed no abnormality, while in four patients there was a 30–50% reduction in tracer uptake. Three of these went on to early cerebral infarction. The authors suggested that persisting focal hypoperfusion after a TIA with full recovery might predispose to the development of early cerebral infarction. Other authors suggested that persistence of hypoperfusion might reflect the duration of the TIA or the severity of associated ipsilateral carotid artery disease. The explanation of persisting hypoperfusion in patients with normal CT scans who make full functional recovery remains obscure and the term 'incomplete infarction' was introduced by Lassen et al.[6] to describe this state. The presence of a persisting area of hypoperfusion on SPECT scanning in patients with TIAs is not difficult to conceptualize; one can postulate reduced blood flow with an increased oxygen extraction allowing preservation of function. Clearly this region is an 'at risk area' and the persistence of hypoperfusion is likely to correlate with a poorer prognosis than that seen in patients in whom flow rapidly returns to normal.

Interest in TIA has extended to the use of specific physiological or pharmacological challenges to emphasize changes in intracranial haemodynamics. Acetazolamide induces cerebral microvascular dilatation. In haemodynamically significant extracranial vascular disease resting cerebral blood flow is maintained by maximal vasodilatation. A further challenge with acetazolamide or CO_2 inhalation cannot be met, so there is negligible change in rCBF in the area affected by stenosis, while other areas vasodilate normally. This allows detection of areas whose resting perfusion is normal but which have effectively no reserve, and identifies patients with extracranial vascular disease who are at high risk of cerebral infarction. It has also been suggested that such challenge tests might be used to identify patients with extracranial vascular disease, but with the availability of ultrasound techniques for this, SPECT as a screening investigation would seem unnecessary. Chollet et al.[7] used acetazolamide as a vasodilator. One gram of this drug was given intravenously to 15 patients who had experienced a TIA in the internal carotid artery territory within the preceding month. Cerebral blood flow was measured using intravenous ^{133}Xe SPECT: two CBF tomograms were obtained from each patient, the first at rest and the second 1–4h later, 10min after the intravenous injection. A control group underwent two scans but did not receive acetazolamide. All patients either also underwent arteriography, intravenous angiography or carotid Doppler ultrasonography with spectral analysis and echo tomography. Focal abnormalities were detected by visual inspection and calculations of inter-hemispheric asymmetry, made before and after acetazolamide administration. Reactivity to acetazolamide was approximately 40% or more in most patients. Seven patients exhibited focal hypoperfusion at rest in the symptomatic hemisphere, this becoming more evident after the administration of acetazolamide, and seven further patients with normal resting CBF tomograms exhibited hyporeactive areas after acetazolamide administration. Thus abnormalities were detected in 14 of 15 patients studied. In nine patients with no significant lesion of the internal carotid artery the areas of hypoperfusion were small. In six patients with severe stenotic lesions of the internal carotid artery, larger and more pronounced changes were noted. It was felt that the discrepancy between the two groups could be ascribed to the haemodynamic influence of internal carotid artery lesions and thus the authors suggested that assessment of cerebral blood flow before and after an acetazolamide challenge in patients with TIA provides insight into the pathophysiology of TIA and possibly identifies patients at greatest risk of progressing to cerebral infarction. Unfortunately the numbers in this study were small, but the premise that the more severe the extracranial vascular disease, the more likely the finding of impaired cerebral vascular reactivity is a reasonable one and supported by our own early work using Xe inhalation cerebral blood flow before and after 5% CO_2 inhalation in patients with graded degrees of occlusive and stenotic unilateral and bilateral carotid artery disease[7,8].

It is also possible that under-responsiveness to vasodilator challenge might reflect chronic damage to the microvasculature as a result of hypertension or ischaemia. In Chollet's study, asymmetry of response was most marked in patients with internal carotid artery disease and was slight when this was absent. Vasoparesis was felt to be the likely mechanism. In patients with extracranial vascular disease, asymmetries were present before acetazolamide and exaggerated afterwards. In persons without significant extracranial vascular disease pre-acetazolamide scans showed symmetry, post-acetazolamide scans only mild asymmetry. A more recent study[9] using [^{99m}Tc]HMPAO SPECT with the vasodilator Ibudilast and image subtraction techniques in 10 patients with TIAs has confirmed the findings of Chollet et al.[7].

The results of these studies in patients with TIA with occlusive or stenotic extracranial vascular disease are of pathophysiological relevance but could hardly support the use of SPECT as a detector of carotid stenosis when ultrasound of extracranial vessels is so freely available, and is obviously a much more direct test. SPECT might, however, be used to monitor the effect of surgical treatment of TIA (carotid endarterectomy) on cerebral perfusion. This will be discussed later. Figure 4.3 illustrates the importance of considering SPECT results in relation to other imaging results.

SPECT in cerebral infarction

Since the late 1960s in the USA there has been a gradual decline in stroke mortality, recently by as much as 5% per annum. There is evidence that these figures reflect risk factor control, better treatment and improved diagnostic criteria, rather than a truly declining incidence. The national hospital discharge survey from the early 1970s to the late 1980s shows a small rise in admissions of patients suffering from cerebrovascular disease, but a decline in death rate from 26% to 17%. These figures contradict the nihilistic approach to acute management adopted by some, and can possibly be improved upon further with designated acute stroke services and access to trials of novel therapies such as neuroprotective and thrombolytic agents.

Stroke can be classified into specific subtypes based upon clinical and radiological criteria. An immediate distinction based on underlying pathophysiology separates stroke into infarction and haemorrhage: infarction accounts for 70–80% of all strokes, intracranial haemorrhage accounting for the rest. Clinical determinants alone are often inadequate for proper classification, and laboratory investigations, in particular structural imaging, are necessary for confirmation of stroke type. In cerebral infarction, regional cerebral blood flow changes occur simultaneously with clinical onset. CT may remain normal for some hours, or even days. MRI diffusion techniques do show immediate ischaemic changes, but more available estimations of changes in T2 signal as a consequence of increase in water content may take up to 8 h to develop. These differences in sensitivity between structural and functional imaging disappear at approximately 72 h.

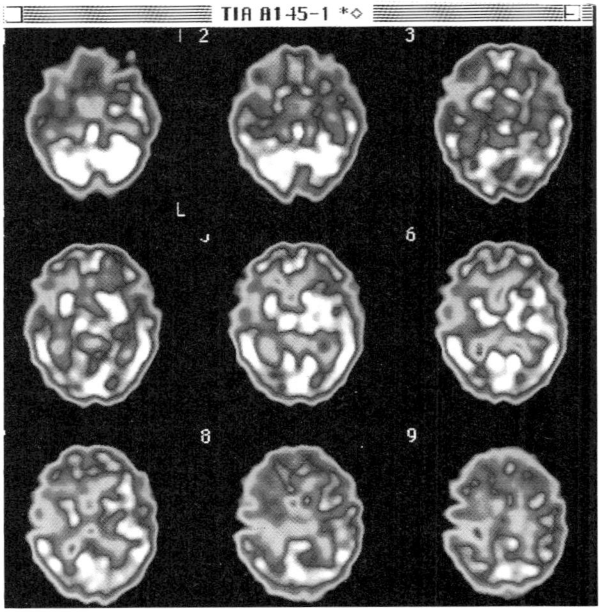

Figure 4.3. HMPAO SPECT scan of a patient with a history of right hemisphere TIA occurring 2 months previously. Doppler ultrasound indicated a right internal carotid artery stenosis and left internal carotid artery occlusion. Cranial CT showed no evidence of infarction, and cerebral angiography confirmed a high grade '80%' right carotid artery stenosis with left internal carotid artery occlusion. The SPECT scan shows cerebral regional blood flow maintained on the left, ipsilateral to the occlusion, but reduced on the right on the side of the carotid stenosis in the territory of the middle cerebral artery. These changes imply an interhemispheric steal and indicate the complexity of assessing cerebral blood flow in patients without knowledge of their carotid artery status, a fault implicit in all early studies of cerebral blood flow change in TIAs.

Early imaging of regional cerebral blood flow by SPECT may therefore allow immediate determination of site and mechanism of infarction in an individual patient. The identification of patterns of hypoperfusion could for instance separate territorial (vascular territory) from non-territorial (haemodynamic) infarction[10].

SPECT appearances in cerebral infarction alter with the passage of time and can be divided into acute, sub-acute and chronic phases. In the acute phase 90% of rCBF SPECT scans will show an area of hypoperfusion, the size being of probable prognostic significance and the distribution of abnormality correlating with eventual CT lesion (although the latter is often small in comparison with the initial area or volume of hypoperfused brain). These changes usually remain unaltered for the first 72 h, although reperfusion can occur within this phase with restitution of near-normal blood flow; the presence and degree of reperfusion often correlated well with eventual clinical outcome. During this acute phase hypoperfusion may affect the whole vascular territory of a particular vessel, only part of which eventually shows structural change. For example subcortical

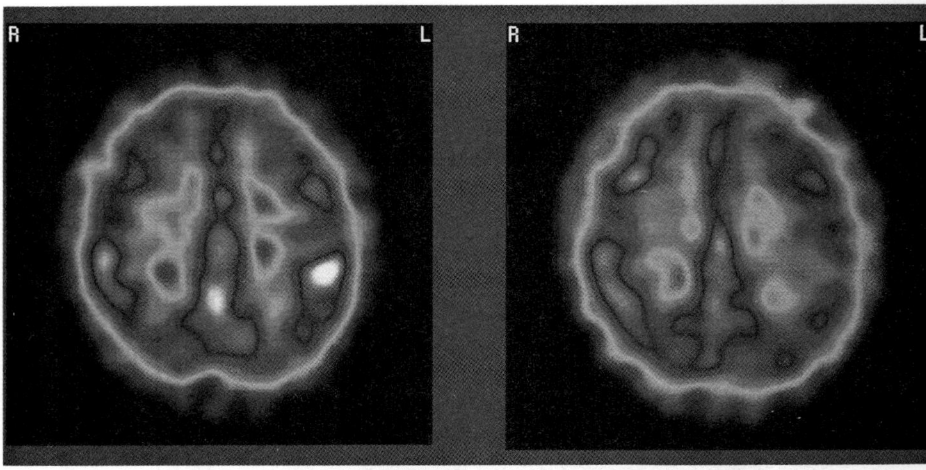

Figure 4.4. A 30-year-old man suffered a parieto-occipital stroke, with visual field defect. This high axial slice shows luxury perfusion at the anatomic site of the lesion at 7 days, which has resolved at 6 months. (For colour version of this figure see page 184.)

infarction in the territory of a perforating ventriculo-striate branch of the middle cerebral artery may show an initial perfusion defect involving the whole of the middle cerebral artery territory. Hypoperfusion may also be seen at a distance from the primary affected area (diaschisis). In the acute phase such rCBF changes contralateral to the site of ischaemia can result in large regions of decrease. An area of reduced rCBF in the unaffected hemisphere usually corresponds with a comparable vascular territory within the infarcted hemisphere.

In the sub-acute phase which may be of variable duration (between 3 and 50 days), the sensitivity of CT and SPECT as diagnostic and outcome measures are nearly identical. At this stage CT begins to show a region of low attenuation which correlates with the pathological infarct volume. MRI similarly will show a hyperintense abnormality on T2 weighted images, and with contrast enhancement will confirm breakdown of the blood–brain barrier, with a relative increase in cerebral blood volume. On SPECT imaging two areas can now be defined, a central zone of severe hypoperfusion and a peripheral zone of lesser hypoperfusion, both of which usually lie within the infarct as defined by CT. The central area initially continues to show hypoperfusion: this may gradually normalize with respect to contralateral hemisphere with the passage of time. HMPAO studies may detect a zone or region of luxury perfusion (defined as flow in excess to metabolic requirements), which lies at the periphery of the affected area (Figure 4.4, see also colour section). This is less evident using IMP as a tracer, however, as pH effects reduce retention. The peripheral zone may then sometimes show a reduction in flow, but is able to respond to the vasodilatory challenge of carbon dioxide or acetazolamide. It is suggested that this peripheral area may represent the ischaemic penumbra or local deafferented or disconnected tissue. Contralateral transhemispheric changes are commonly seen sub-

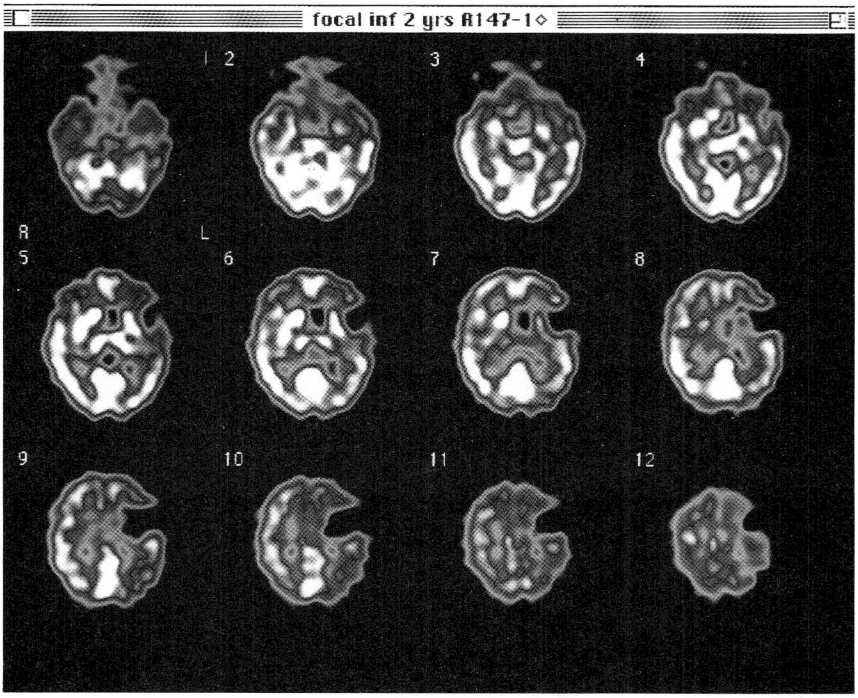

Figure 4.5. A 56-year-old woman who had undergone coronary artery bypass grafting developed atrial fibrillation and had bilateral carotid artery disease with an established infarct in the left hemisphere some 2 years previously from which a good functional recovery had been made. SPECT imaging shows an area of marked hypoperfusion in the left MCA territory. CT scans confirm that this abnormality reflects tissue loss.

acutely, but less so than during the acute phase. Studies with ^{133}Xe PET and SPECT have shown that the hypoperfusion and hypometabolism seen in the contralateral hemisphere gradually returns to normal with the passage of time, and does not persist into the chronic phase.

Finally, in the chronic phase, CT shows an area of hypodensity while MRI shows an increased T2 signal which does not enhance with contrast. SPECT shows an area of non-perfusion consistent with loss of brain tissue. In subcortical areas ipsilateral to a cortical infarct, reductions of 30–40% in blood flow can be seen (Figure 4.5). SPECT abnormalities are less obvious in small subcortical lesions and in diffuse periventricular established vascular disease.

Raynaud et al.[11] used IMP SPECT to assess 16 patients during acute, subacute and chronic phases of cerebral infarction. Their patients had suffered a mixture of cortical and subcortical events, though the majority had a cortical component. They were able to confirm the presence of central and peripheral zones of infarction and demonstrated changes in the contralateral hemisphere. Using acetazolamide to assess vascular reactivity, they concluded that the central area represented the infarct area and the peripheral area the peri-infarct zone or penumbra. The outcome in these two zones, as defined in the subacute phase, appeared to be different. IMP uptake at 10 min, 5 h and 24 h in the central zone was compared with comparable changes in the peripheral zone: they concluded that the central zone corresponded to necrosis, while the peripheral region was often eventually spared from structural change, as defined on cranial CT. They termed the peripheral zone 'the zone of salvage' and suggested that knowledge of the existence and extent of these changes in the peripheral zone might facilitate the selection of patients with cerebral infarction for new treatments. They agreed, however, that influence of pH on IMP tracer uptake clearly would confound and confuse interpretation. The same authors in a subsequent publication[12] proceeded to a pathophysiological study of chronic infarction with IMP to evaluate the relevance of the peri-infarct zone. Seventeen patients were studied with ^{133}Xe and IMP, uptake being assessed at 10 min, 2 h and 5 h. Scan timing varied considerably from 3 to 16 months after infarction. All infarctions were cortical, and PET was used to assess rCBF in some patients. Two zones were again defined. The central zone showed a severe decrease in IMP uptake, correlating with low attenuation on CT, while the peri-infarct or peripheral zone showed a moderate decrease in IMP uptake, this latter region having a normal appearance on CT. The zone of moderately reduced IMP uptake was not seen in early scans, but was evident at 5 h: the suggestion was made that the peripheral or peri-infarct region represented not neuronal loss but deafferentation. Interestingly, the volume of this zone related to the neurological status of the patients. This study can obviously be criticized for the small number of patients evaluated and in that a high proportion of patients had severe carotid artery stenosis or occlusion, which might alter autoregulatory or vasoreactive changes in the peri-infarct zone, potentially confounding interpretation of the abnormalities seen, certainly in the chronic phase. De Roo et al.[13] assessed 92 patients with infarction and 30 with TIA using HMPAO and confirmed most of Raynaud's findings with regard to the zones of central and peripheral or peri-infarct change. They were also able to demonstrate hypoperfusion remote from the infarct, due to diaschisis.

CT and SPECT have comparable sensitivity in the subacute phase of cerebral infarction, with detection rates of 89.5% and 85.7% respectively. False-negative results have been reported with SPECT in lacunar and other subcortical infarcts, presumably due to lack of spatial resolution. Cordes et al.[14] compared HMPAO SPECT and MRI with gadolinium enhancement in 19 subacute and seven chronic cases. They also estimated regional cerebral blood volume. They found that all subacute lesions showed an increase in cerebral blood volume, and that rCBF could be either increased or decreased. Contrast enhancement on MRI was generally associated with an increase in regional cerebral blood volume, but not with increased rCBF. In all chronic lesions studied, both regional cerebral blood volume and blood flow decreased and there was no evidence of contrast enhancement detectable by MRI. They concluded that SPECT revealed an increase in regional cerebral blood volume indicating vasodilatation or vasoparalysis and that rCBF was either increased or decreased, depending upon collateral circulation or reperfusion in involved vascular territory. It is clear from this study that just looking at rCBF in isolation may be misleading and that other aspects of cerebrovascular haemodynamics, such as cerebral blood volume, must be taken into account if proper physiological interpretation of data is to be made. More recently Baird et al.[15] using [^{99m}Tc]HMPAO SPECT studied 35 acute ischaemic strokes, the scans being performed during the acute, subacute and chronic phase. The study was somewhat incomplete in that only nine patients underwent scanning in all three phases of the evolution of their infarct. The majority of infarcts were in the middle cerebral artery territory and a small proportion were in the posterior circulation; in this study too there was a mixture of cortical and subcortical infarction. Their study was of interest in that hyperaemia or reperfusion in the involved vascular territory was identified in 10 of the 15 patients who underwent two scans. Being evident in the subacute phase, these changes confirmed the findings of Raynaud, but late SPECT scans showed perfusion defects greater than the eventual infarct size as defined on CT. They also demonstrated crossed cerebellar diaschisis in eight patients with hemisphere infarction and cortical–subcortical diaschisis in one.

All early rCBF SPECT studies have addressed the question of sensitivity when SPECT is compared with CT or MRI, and it is generally accepted that, with the exception of sophisticated MRI diffusion techniques, early rCBF SPECT is the most sensitive confirmatory investigation for the diagnosis of cerebral infarction. In addition, the location of a perfusion defect may help separate thromboembolic or territorial cerebral infarction from haemodynamic or non-territorial ischaemic events. SPECT will not replace CT as the initial routine investigation, as early separation of haemorrhage from infarction is dependent upon structural rather than functional imaging. The changes defined by various workers through acute, subacute and chronic phases are well accepted, though the numbers of patients studied have been small and the studies in themselves have often been incomplete, and contain a heterogeneous group of infarct subtypes. It would appear from the available literature, however, that SPECT can confirm initial diagnosis, and that subsequent patterns of flow may reflect eventual clinical outcome. It is potentially an important investigation in the selection of patients for and the monitoring of novel therapies. Whether SPECT is sensitive enough to confirm the presence and extent of 'the zone of salvage' remains to be proven, and may depend on the use of high resolution imagers.

SPECT as a predictor of neurological recovery in cerebral infarction

Prediction of outcome following cerebral infarction is uncertain. Clinical parameters such as age, the degree of motor deficit, the presence of gaze palsy, impairment of conscious level and the presence of incontinence, will give rough guides as to outcome, as will location and size of the eventual lesion as demonstrated on cranial CT. With the advent of new therapies, consideration has been given to the use of rCBF measurements as an index of eventual outcome. Such studies have been hampered by inconsistency of timing of scans, as well as their application to groups of stroke patients in whom a good outcome could be in any case anticipated, such as subcortical or lacunar infarction and posterior circulation cortical infarction. Lee et al.[16] used IMP SPECT to study 16 patients with acute cerebral infarction, all of whom had signs and symptoms referrable to middle cerebral artery territories. Images were acquired between 20 and 30 min after [^{123}I]IMP injection. Regions of interest were compared with CT defined infarction and count densities indicative of tracer retention were compared with those in equivalent regions in the contralateral hemisphere. Patients were classified according to degree of improvement and long-term outcome. The patient with the greatest deficit generally had the largest CT lesion and the greatest reduction (up to 80%) in isotope uptake. However, correlation between count density, anatomical lesion size and outcome was generally poor. This study did not document the volume of the perfusion defect, the acquisition obtained limited axial slices, and time from infarct to evaluation was highly variable, but it nonetheless provided a starting point for similar studies. Bushnell et al.[17] evaluated 14 patients with major anterior circulation infarction, using IMP. All scans were performed within 30 days of onset and IMP defect volume was calculated and expressed as a fraction of the volume of the whole ipsilateral hemisphere. The methodology was somewhat unusual in that scans were read after division into 12 equal radial sections and a scoring system was introduced to compare counts within each comparable zone, comparing affected and unaffected hemisphere. The magnitude of flow disturbance appeared to correlate with early death but not with Barthel index in the 20 survivors. The authors emphasized the value of SPECT immediately after ictus in identifying patients who might die early from tonsillar or transtentorial herniation. They noted that, in contrast to scans carried out early and at 6 months, scans at 2 weeks did not appear to correlate with outcome.

A more comprehensive study by Giubilei et al.[18] investigated 32 patients with completed ischaemic stroke within 6 h of the event, using CT, cerebral angiography and [^{99m}Tc]HMPAO SPECT. Follow-up CT and rCBF studies were performed at one week and one month. All 10 patients with early severe impairment of both neurological status and rCBF had poor outcome at one month, but blood flow changes in the subacute phase did not correlate well with eventual outcome. They concluded that early evaluation of rCBF with SPECT is useful in detecting subgroups of patients with adverse clinical outcome.

Studies using PET of perfusion and cerebral metabolic rate of oxygen ($CMRO_2$) during the first 18 h of stroke have strongly suggested that reperfusion correlates with improved prognosis. Baird et al.[19] demonstrated the prognostic value of reperfusion during the first 48 h of ischaemic stroke in an important HMPAO SPECT study. Each patient on admission (mean 8.2 h from ictus) had SPECT imaging, with a repeat study approximately 24 h after the first. The volume of defect on each scan was measured, and at least 50% of the patients studied showed a degree of partial reperfusion, this being defined as a decrease of deficit volume of 30% when compared with the initial scan. Patients with this degree of reperfusion appeared to have significantly smaller infarcts on CT performed some 7–10 days later, as well as lower mortality and better functional outcome at 3 months. The authors concluded that early reperfusion appears to be of prognostic value and demonstrated that this can be assessed by using SPECT. They suggested that serial studies allowing semiquantification of reperfusion could be used as an index of tissue salvage and might be useful in evaluating new therapies. They also demonstrated that natural reperfusion was not detrimental to outcome. This observation is important, as it had been suggested that drug-induced thrombolysis and reperfusion might provoke cerebral oedema or be associated with haemorrhagic transformation of infarct (even though this was not borne out by early safety and efficacy studies).

As a prognostic guide, SPECT appeared initially disappointing, its evaluation being hampered by small incomplete studies that had looked at heterogeneous groups of patients in whom the time between ictus and initial scan was highly variable, and by non-uniform methodology. It has been shown that early SPECT is a useful diagnostic tool in cerebral infarction; it may also be a useful prognostic investigation and serial studies may allow estimation of the 'window of efficacy' for new treatments. Rossell et al.[20] have attempted, for instance, to evaluate the therapeutic effectiveness of nimodipine in acute focal cerebral ischaemia, comparing a clinical evaluation scale with flowmetric evaluation, SPECT being performed serially during the course of treatment. Although the study in itself was inconclusive and the role of nimodipine in acute infarction appears not to have been established, this paper shows how simple blood flow estimations can be made in tandem with clinical trials of treatment.

Diaschisis

An occasional finding in PET studies of patients with cerebrovascular disease has been that of decreased cerebral blood flow or metabolism in structurally normal brain at a distance from the area of infarction. These findings had been suggested by early cerebral blood flow work. Kempinsky[21] ascribed such changes to interruption of input from afferent connections and reintroduced the term 'diaschisis' originally used by Von Monakow in 1914 to describe spread within the nervous system across the corpus callosum and along intrahemispheric association pathways. These changes 'at a distance' have been documented in anterior hemisphere lesions, where there appears to be alteration in blood flow and metabolism in the opposite cerebellar hemisphere (Figure 4.6). It has also been described in the visual cortex in the context of anterior visual pathway disease and in the cerebral cortex overlying subcortical lesions. Similar changes in the deep nuclei, in particular the thalamus, have been documented in patients with cortical infarction. As mentioned earlier, changes in the contralateral hemisphere have occasionally been reported in cortical disease, mirroring the affected hemisphere. Hemispheric diaschisis is most obvious within the first

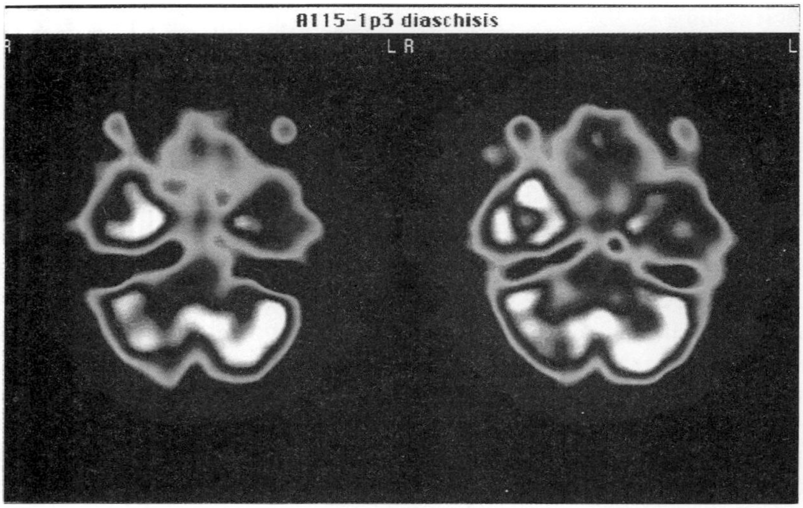

Figure 4.6. SPECT scan of a 36-year-old woman on the oral contraceptive pill. She developed a left middle cerebral artery territory infarct as demonstrated by hypoperfusion shown in these images in the temporal lobe. This scan performed at 10 days post-ictus shows reduced flow in the contralateral cerebellar hemisphere.

few days of infarction. Crossed cerebellar diaschisis was first reported using PET by Baron et al.[22], who noted depressed rCBF and $CMRO_2$ in the contralateral cerebellar hemisphere up to 2 months after anterior hemisphere infarction. The clinical significance of these distant findings remains uncertain, and disturbance of rCBF within a single vascular territory may not reflect true diaschisis, but simply indicate that the initial perfusion defect was more substantial than the resultant infarction, as confirmed by imaging or autopsy. SPECT imaging with radiolabelled tracers would seem an ideal method for investigating the relevance of rCBF changes at distance from infarction, bearing in mind that only high resolution systems are capable of adequately visualizing subcortical structures. Bowler et al.[23] assessed 60 consecutive cerebral infarcts, 18 of which were subcortical. HMPAO SPECT was performed in these patients as soon as possible within the first week and at 3 months. The patients underwent an extensive cerebrovascular examination, including neuropsychometry and CT. Areas of decreased cortical uptake were identified and confirmed as well as quantified by measuring count density in regions of interest and noting percentage changes compared with the unaffected contralateral cerebral hemisphere. In essence, this extensive study showed no significant correlation between the presence of diaschisis and clinical status. Specifically, cortical hypoperfusion did not appear to correlate with any specific neuropsychological deficit. It was concluded that diaschisis was not of clinical relevance. Conversely Bogousslavsky et al.[24] examined in detail two patients with infarction in the territory of the right anterior choroidal artery. These patients had hemiparesis, hemianaesthesia, hemianopia and severe neglect. IMP SPECT showed marked hypoperfusion, not just confined to the posterior capsular region but also involving frontal and parietal cortex. Autopsy in one patient showed infarction limited only to the deep white matter of the internal capsule, and disconnection was thus postulated with ipsilateral deactivation of the parietal frontal cortex and associated neuropsychological deficit. The conflict between these observations is difficult to resolve. Clearly diaschisis can be demonstrated and it is tempting to offer it as an explanation when apparent cognitive dysfunction of a cortical nature occurs in a patient in whom structural imaging demonstrates the ischaemic process to lie in a subcortical site. Indeed syndromes such as thalamic neglect might well be explained on the basis of cortical deafferentation: this is certainly an area which could be further explored. Although the resolution of SPECT may turn out to be a limiting factor, more precise techniques such as PET do not lend themselves easily to the study of acute events in large cohorts of patients.

SPECT and surgery of atherosclerotic cerebrovascular disease

Atherosclerosis has a predilection for the origin and sites of bifurcation of large arteries. The internal carotid artery distal to the carotid bifurcation is a common site for atheroma, with resultant stenosis and eventual occlusion. Carotid endarterectomy (CEA) and extracranial–intracranial bypass (EC–IC bypass) are procedures that have been performed in symptomatic carotid artery stenosis and occlusion, respectively. Large multicentre randomized trials have shown benefit of endarterectomy in high grade carotid artery stenosis, but no benefit has been demonstrated in moderate (<70%) stenosis. Medical treatment appears superior to surgery in patients with <30% stenosis. In carotid artery occlusion, extracranial intracranial bypass has been shown to be ineffective in reducing the risk of infarction.

Carotid endarterectomy

This surgical procedure was introduced in 1954 and in the 1980s became the most common vascular surgical procedure carried out in North America. The implication that the more haemodynamically significant the stenosis is the greater the indication for surgery is supported by several studies. Our own work in assessing cerebrovascular reactivity in patients with carotid artery disease confirms that haemodynamic instability is more commonly found in patients with high grade stenosis or bilateral disease, and is evident in those with a worse prognosis as well as those likely to benefit from surgery[8].

The haemodynamic significance of carotid stenosis can be measured by assessing resting rCBF, and then reassessing after a challenge with a vasodilatory agent such as acetazolamide or 5% CO_2 inhalation. Physiological challenge by alteration in posture may also be used to assess perfusion in patients with carotid artery disease and haemodynamic instability, and thus identify those who might benefit most from surgery. It can also identify those in whom the risk of ischaemic events during surgery is high, allowing modification of surgical technique. The combined morbidity and mortality rates from carotid endarterectomy vary enormously, from 2 to 15%. The risk associated with surgery is influenced by many factors, including the severity of extracranial vascular disease. Surgical techniques vary; the use of indwelling shunts and the benefits of local as opposed to general anaesthesia have their protagonists. Operative monitoring of cerebral function in the form of electroencephalography, evoked potential recordings and transcranial Doppler all have their advocates in protecting from and predicting the ischaemic complications of carotid cross-clamping.

Maurer et al.[25] assessed changes in cerebral profusion before and after carotid endarterectomy using IMP SPECT. Using visual as well as numerical interpretation, they demonstrated an improvement in perfusion of >2 standard deviations ipsilateral to surgery in 15 of 19 patients studied and on the contralateral side in 18. They concluded that quantification of redistribution of blood flow following surgery could provide an objective index of improved perfusion. They also suggested that this technique was useful in detecting changes in the contralateral hemisphere. Hayashida et al.[26] used a supine and erect stress technique with HMPAO to demonstrate impaired autoregulation in a small number of patients with known carotid occlusion, but did not apply this method to postoperative assessment after EC–IC bypass. Algotssom et al.[27] assessed the effect of carotid cross-clamping intraoperatively in 13 patients undergoing endarterectomy. They used a xenon intravenous blood flow technique with 3 min clearance in conjunction with HMPAO SPECT. Patients were anaesthetized with isoflurane and fentanyl; preoperative and two intraoperative measures made before and after carotid artery clamping showed that the degree of decrease in regional cerebral blood flow during cross-clamping correlated strongly with risk of operative ischaemic morbidity. It would seem from these limited studies that SPECT can be used in the same way that EEG monitoring and cortical evoked response measurements are used in the assessment of patients who are to undergo carotid artery surgery, whether this be endarterectomy or occlusion (Figures 4.7, 4.8).

The problem with SPECT imaging in this situation is that one cannot directly quantify the changes in rCBF which occur on test occlusion. Percentage change in isotope uptake may be used for this purpose, however, and our own experience suggests that a reduction of less than 25% in isotope uptake on test occlusion predicts that the patient will tolerate operative occlusion without ischaemic complications. The figure for percentage reduction in uptake is arrived at by comparing asymmetry indices in the same areas on occluded and non-occluded images. Clearly there is the prospect of intraoperative SPECT being of benefit in this situation but it needs in the first instance to be assessed in conjunction with an established technique.

SPECT has been used to assess and monitor the outcome of EC–IC bypass. A randomized trial (EC–IC Bypass Study Group 1985; see ref. 28) demonstrated that this type of surgery did not offer superior protection to medical treatment for these carotid or middle cerebral artery pathologies. However, bypass surgery is still offered for non-atheromatous disease such as moya moya, and some protag-

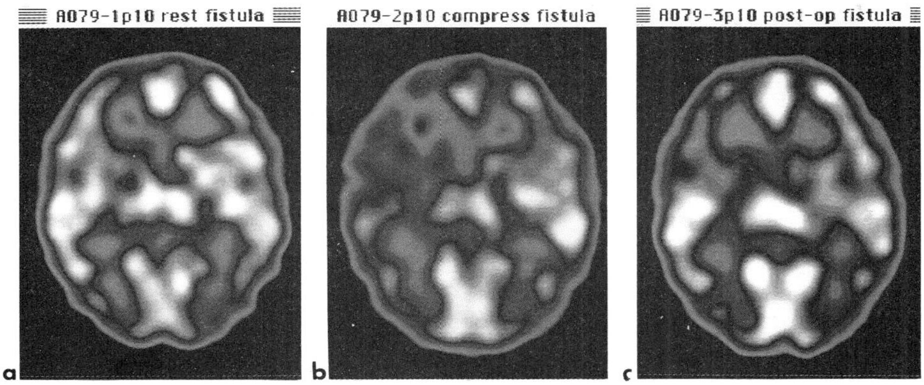

Figure 4.7. SPECT imaging in a patient with a carotico–cavernous fistula. SPECT is carried out (a) at rest, (b) during carotid compression and (c) post-operatively following balloon occlusion of the fistula. A marked reduction in flow is seen during compression, and a lesser reduction post-operatively. These studies may be useful in predicting whether a patient can tolerate such a surgical procedure without significant ischaemic morbidity.

onists still hold that the methodology of the EC–IC study was unfavourably biased against surgery.

Di Piero et al.[28] used HIPDM SPECT to assess 14 patients with carotid occlusion undergoing EC–IC bypass. Prior to surgery all had areas of hypoperfusion in the relevant area of the cerebral cortex. rCBF was improved by the procedure in six. However further scans at 6 and 12 months in conjunction with angiography to confirm patency of the bypass, showed that rCBF had reverted to its preoperative state. They concluded that the operation had failed to produce any long-term benefit in cerebral perfusion. Batjer et al.[29] used ^{133}Xe inhalation SPECT in patients who had developed focal neurological deficit after carotid occlusion. They measured resting blood flow and carried out vasoreactivity studies after intravenous administration of acetazolamide before and at 1 and 9 weeks after EC–IC bypass. This study not only confirmed improvement in resting flow but also demonstrated a more significant improvement in vasoreactivity. The authors emphasized that the improvement in cerebral vascular reserve as detected by activation studies was a more important physiological parameter of benefit from surgery than resting blood flow. Neither of these studies, however, fully answers the question: Di Piero did not assess cerebrovascular reactivity and Batjer's follow-up was too short to counter Di Piero's criticism that benefits were only short-term. Unless EC–IC bypass emerges again as a widely performed surgical procedure it seems unlikely that the application of SPECT to assess physiological benefit will be further explored. Nonetheless this surgical procedure is carried out for rare cases of obliterative cerebrovascular disease such as moya moya and also in patients who might be about to undergo carotid artery occlusion electively in the management of intracavernous aneurysm or a slowly progressive carotid occlusion from skull-based tumour. In such cases, resting and activation SPECT imaging can be used not only to confirm the physiological

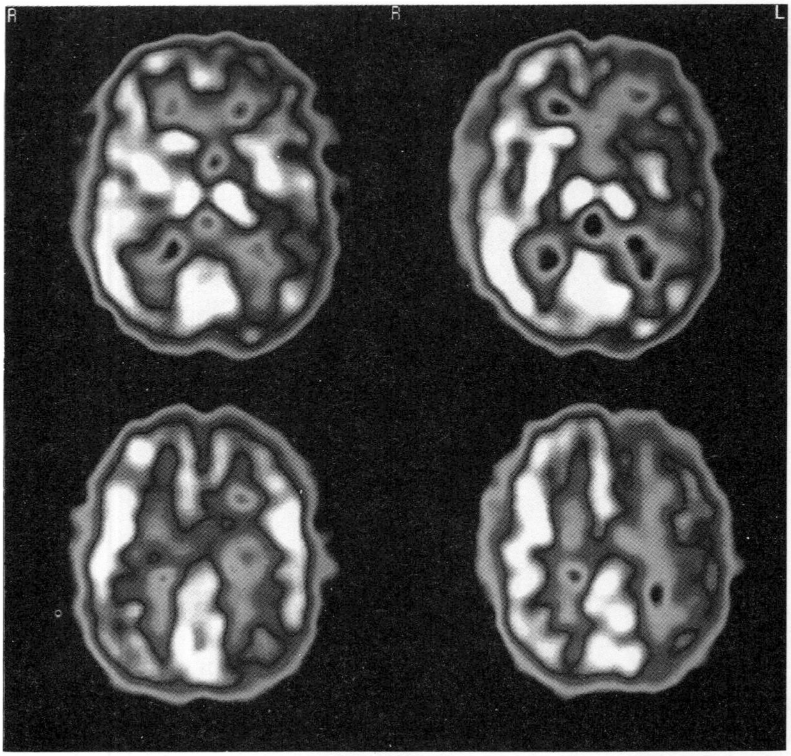

Figure 4.8. A patient with a giant aneurysm of the intracranial portion of the left internal carotid artery. Blood flow measurements were made before (left) and after (right) compression of the carotid artery to evaluate whether the patient would be able to tolerate luminal occlusion.

benefit of the procedure, but to select patients and determine the timing of surgery by detecting areas of reduced vascular reserve.

SPECT in migraine and other headache disorders

The incidence of migraine is thought to be as high as 19% of males and 29% of females. Assessment of lifetime events indicates approximately 60 million sufferers in the USA alone. Migraine creates a considerable economic burden, accounting for 3–5% of all time loss from employment in the UK, despite the fact that two-thirds of all attacks occur in non-working hours. In recent years the International Headache Society (IHS) has introduced a simplified classification. Migraine without aura is manifest by unilateral throbbing headache which is moderate to severe in intensity, aggravated or precipitated by exercise, and which is associated with nausea, photophobia and phonophobia. Migraine with aura is defined by similar symptoms which are, in addition, associated with retinal, cortical or brain stem signs of dysfunction, these rarely being longer than

60 min in duration and generally preceding the onset of headache disorder. There are both vascular and neurogenic hypotheses of the pathophysiology of migraine, and although a combination of the two seems likely it is accepted that cerebral blood vessels play an important if not central role in the development of migraine. Heyck[30] suggested an opening of carotid arteriovenous connections, triggered by serotonergic mechanisms, resulting in diversion of blood from the capillary bed with resultant ischaemia. Cerebral blood vessels receive sensory innervation through trigeminal vascular pathways, and vasoactive neurotransmitters of central and peripheral origin may act upon them, resulting in vasodilatation and pain. Biochemical, mechanical and immunological factors may modify this pathway. rCBF studies in migraine have been logistically difficult to perform because of the short duration of aura and unpredictability of acute migraine attacks. Migraine with aura has been precipitated by angiography, and this situation has occasionally been used for an assessment of rCBF, showing reduced rCBF during the attack, which reaches an ischaemic threshold during the aura phase. This abnormality of rCBF often persists after the aura has passed. The duration of hypoperfusion in migraine has not been fully studied. Anderson et al.[31] carried out repeated rCBF studies in patients with migraine with aura to assess the temporal pattern of blood flow changes. They studied several patients, measuring rCBF by ^{133}Xe inhalation with SPECT. Mean flows in regions of interest were calculated and compared with homotopic regions in the opposite cerebral hemisphere. Focal symptoms of hemisensory or hemivisual type were noted and anatomically localized. Duration of headache varied between 3 and 26 h and frequency of attack between 2 and 6 per year. rCBF measurements were made within 2–6 h of onset, at 24 h and then finally at one week. Initial hypoperfusion was followed by early hyperperfusion. Perfusion at 1 week was symmetrical. The methodology was complex but the results were remarkably uniform and seemed to demonstrate that the onset of headache often occurred during the phase of hypoperfusion rather than, as one might have anticipated, during hyperperfusion. Indeed the change from hypo- to hyperperfusion often coincided with gradual regression in the headache. Flow defects were all unilateral and their location did appear to correlate with the clinical localization of symptoms. These findings were essentially confined to the cortex, occasionally extending into white matter. The mean decrease in rCBF was, however, relatively small as was the increase during the headache phase. The authors concluded that the transition from hypo- to hyperperfusion did not coincide with the onset of headache, thus contradicting some theories regarding the vascular mechanism of headache in migraine. This study also suggested that hypoperfusion could persist longer than was generally thought. An interesting finding from the study was that the degree of reduction in rCBF did not appear to approach the ischaemic threshold: previous workers had suggested that this was the case, citing it as evidence that the focal neurological symptoms of migraine had an ischaemic basis.

A more recent study by Schlake et al.[32] addressed the question of rCBF changes occurring in patients between migraine attacks. They used HMPAO

SPECT to investigate a heterogeneous group of 33 patients, the majority of whom had migraine with aura, a minority with migraine without aura, and some with cluster headache. Regional hypoperfusion closely related to the site of hemisphere neurological symptoms associated with migraine attacks was seen in 17 patients. The patients with cluster headache had normal rCBF. The authors concluded that interictal hypoperfusion might be a common finding in migraine with aura, and that the attacks in themselves might represent an exacerbation of a subclinical disturbance of the microvasculature. This suggests the possibility that SPECT might have some predictive value in identifying the rare patient who develops migraine stroke, and certainly in a patient who demonstrates persisting hypoperfusion between migraine attacks, one might wish to avoid exposure to procoagulants such as the oral contraceptive. This requires further evaluation. A further area of interest would be to assess the response in patients with persisting hypoperfusion to vasoactive challenge with acetazolamide or carbon dioxide. This would give some evidence as to whether there was a persisting disorder of cerebrovascular reactivity.

Transient global amnesia

This unusual condition presents as an episode of total amnesia of some hours' duration followed by complete recovery of memory, except for events that occurred during the episode itself. While symptomatic, a patient will often answer simple questions and appears otherwise normal, and may carry out complex tasks such as driving a car. Neurological examination remains normal. The pathophysiological disturbance associated with transient global amnesia remains conjectural but presumably involves both temporal lobes simultaneously. Seizure disorders, migraine variants and vascular insufficiency of an atherosclerotic nature have all been postulated as causes, and indeed all may, in different individuals, account for the clinical presentation. However studies of the prevalence of vascular risk factors in sufferers show no significant difference from those present in an age- and sex-matched control population. Physiological studies with PET and EEG have suggested a reduction in cerebral blood flow and metabolism as well as a disturbance of electrical function in both temporal lobes at the time of an attack. Stillhard et al.[33] reported a single case in whom neuropsychological assessment and SPECT imaging with HMPAO were undertaken during and after the event. During this episode patchy retrograde amnesia was noted, with difficulty in learning both new, verbal and non-verbal material. SPECT scanning confirmed severe bitemporal hypoperfusion. Recovery was associated with return to neuropsychological normality, and SPECT imaging at 3 weeks and at one year showed normal flow. This case study seems to confirm a vascular component as underlying transient global amnesia but does not discriminate between vertebrobasilar insufficiency and migraine. The absence of vascular risk factors, the generally benign and relatively non-repetitive nature of this condition, as well as the association of its onset with exercise perhaps points more to vasospastic etiology. This single case report does emphasize the value of SPECT when carried out in the acute phase of an unusual problem in which ischaemia may be a possible underlying mechanism.

Dementia and cerebrovascular disease

The pathophysiological concept of vascular dementia is changing. The old hypothesis that intellectual impairment was a consequence of chronic cerebral hypoperfusion and ischaemia has been abandoned, and the concept of multi-infarct dementia introduced in the 1970s has gained increasing recognition, especially with the improved sensitivity and increased availability of structural imaging. It has also become evident that dementia may result from extensive diffuse white matter ischaemic disease in the absence of large cortical infarcts. This subcortical pathology results from disease of small blood vessels, and is often insidious in its development. In Western countries, vascular disease is the second most important cause of dementia after Alzheimer's disease, accounting for approximately 15% of all cases. The concordance rate between clinical and pathological diagnosis is much lower for vascular dementia than for Alzheimer's disease. The Hachinski ischaemic scale forms the basis of the criteria most widely used for the diagnosis of vascular dementia. It does appear to separate pure forms of Alzheimer's disease from vascular dementia, but has difficulties where the two might co-exist. FDG or O_{15} PET can provide an accurate diagnosis of Alzheimer's disease and can distinguish it from the multi-infarct state. However the lack of availability of PET restricts its widespread clinical application. Cohen et al.[34] have used IMP SPECT in the evaluation of patients with vascular dementia, demonstrating multiple asymmetric defects involving both grey matter and white matter of cerebral hemispheres. They concluded that SPECT imaging with IMP was more sensitive and accurate than CT, and implied that the former could be widely used for diagnostic confirmation. Despite this, functional imaging of blood flow has yet to find a place in currently used diagnostic criteria for vascular dementia and does have an obvious drawback in the detection of small white matter lesions such as those that form the pathological basis of subcortical dementia. Some recent work, however, has suggested that the use of a hypercapnic challenge can improve the sensitivity of rCBF SPECT in this disorder. The role of SPECT in dementia is discussed more fully in Chapter 5 (Figure 4.9).

Subarachnoid haemorrhage

Experimental and clinical studies of CBF following subarachnoid haemorrhage have employed the full range of available techniques. Early non-tomographic xenon techniques allowed serial measurements but provided images of limited spatial resolution and in particular suffered from an inability to identify very low flow areas. Nevertheless, useful information about both early and sequential

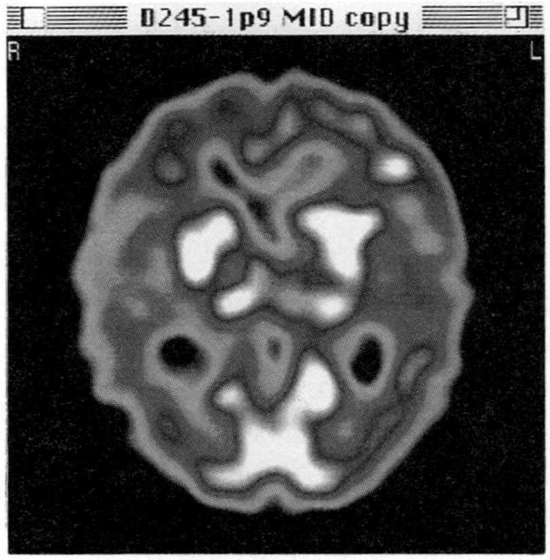

Figure 4.9. A 56-year-old man with a history of hypertension, episodes of focal neurological dysfunction and significant cognitive impairment. SPECT scan shows widespread reduction in cortical blood flow compatible with the diagnosis of a multi-infarct dementia.

changes of global or hemispheral blood flow were obtained. These techniques have, however, been supplanted by tomographic methods. Initially SPECT was used in association with ^{133}Xe[35]. PET studies were also carried out[36,37]. More recently the standard tracers [^{99m}Tc]HMPAO[38–42] and ^{123}I IMP[43] have been used in SPECT studies of subarachnoid haemorrhage.

The major advantage of imaging blood flow as opposed to other parameters in subarachnoid haemorrhage is similar to that in other causes of cerebrovascular ischaemia, namely the early detection of functional changes before abnormalities become visible on structural images. The causes, time course, and degree of reversibility of cerebral ischaemia caused by subarachnoid haemorrhage differ significantly from those in atherothrombotic disease. In particular, there exists the opportunity to study cerebral haemodynamics before the onset of ischaemic symptoms: perfusion changes may occur after the initial bleed but before the onset of delayed ischaemia, the so-called 'second stroke'. This raises the prospect of identifying and treating 'at risk' patients, a prospect which has been a major driving force in producing detailed rCBF studies in the early stages after subarachnoid haemorrhage.

Correlation between areas of abnormal perfusion and clinical progress and outcome, and measurements of the effects of medical and surgical intervention on cerebral blood flow, have all been studied using SPECT with xenon and other tracers. The prospects of monitoring changes in rCBF occurring in association with the use of newer treatments such as angioplasty, thrombolytic and neuroprotective agents, offer an expanding role for CBF studies in subarachnoid haemorrhage.

Cerebral haemodynamics following subarachnoid haemorrhage

Following rupture of a cerebral arterial aneurysm into the subarachnoid space, there is an immediate but short-lived increase in intracranial pressure, and a consequent transient decrease in cerebral blood flow, which is probably global. The critical phase of this response lasts seconds to minutes, making cerebral blood flow studies at this time extremely difficult, but intracranial pressure recordings and transcranial Doppler studies (performed during aneurysm re-rupture) have documented these early cerebral circulatory changes. The delayed process of cerebral vasospasm ensues, caused by cerebral arterial narrowing which develops over several days, and which causes further cerebral blood flow changes. The aetiology of arterial spasm is unknown, but it probably results from vasoactive agents released into the subarachnoid space during the initial bleed.

Documentation of the time course of cerebral vasospasm has been carried out using serial cerebral angiography, transcranial Doppler sonography, and cerebral blood flow studies. Factors other than vasospasm may affect cerebral blood flow; mechanical complications (hydrocephalus, cerebral swelling, surgical manipulation), recurrent haemorrhage, and disturbances of systemic haemodynamics, metabolism, and electrolytes (principally hyponatraemia). It is widely accepted, however, that delayed vasospasm is the major cause of reversible ischaemic damage following subarachnoid haemorrhage[40–44]. The alternative causes of neurological deterioration after subarachnoid haemorrhage can be assessed readily by monitoring of fluid and electrolyte balance, blood pressure, and arterial gas pressures, and by the use of CT to detect rebleeding or hydrocephalus.

The diagnosis of incipient vasospasm is more problematic, as serial angiography to monitor reduced arterial diameters is invasive and potentially damaging, and CT will detect ischaemia only once it is irreversible. Cerebral blood flow studies, and more recently the transcranial Doppler technique which measures arterial flow velocities, have been extensively evaluated, with the aim of identifying patients at risk of delayed ischaemia who might benefit from targeted medical therapies or tailored surgical management.

Early studies of CBF

Early studies of CBF after subarachnoid haemorrhage allowed only a global estimation of flow, usually from each cerebral hemisphere. The earliest studies employed the nitrous oxide method, which was followed by the development of several ^{133}Xe techniques, usually involving intra-arterial injections. Despite their limited spatial resolution[45], these initial studies established that cerebral perfusion following subarachnoid haemorrhage was usually lower than normal. Correlation between CBF and the patient's level of consciousness, and with the presence of angiographic vasospasm was also shown.

A further significant advance came from a large serial study of the time course of CBF, which also allowed a more detailed examination of the relationship between abnormal CBF and various clinical parameters. A total of 1265 measurements were made in the first 3 weeks following subarachnoid haemorrhage in 116 patients[46]. The study included a series of 47 volunteers, allowing control comparisons to be made. From this it was established that CBF falls during the first 2 weeks after subarachnoid haemorrhage, and remains low in the subsequent week. These CBF changes appeared to be independent of the site of the ruptured aneurysm, and were similar in patients in whom angiography had shown an aneurysm and those in whom angiography was normal. Additional clinical and radiological correlations have been broadly confirmed in much of the subsequent work in this area: CBF is lower in patients in a poor clinical grade (and also, more simply, in patients who were drowsy) than in those in a good clinical grade (non-drowsy); CBF is also lower in older than younger patients even when correction for clinical status is made. Additionally, it was found that patients with a full recovery at 3 months have higher and more stable CBF levels than patients with poor outcomes. Age was discovered to be an extremely important determinant of CBF levels: the normal inverse relationship between age and CBF was maintained but exaggerated in subarachnoid haemorrhage[46], so that the margin between CBF levels and the ischaemic threshold was lowest in elderly patients, putting them at the greatest risk of complication from other adverse influences on cerebral perfusion, such as arterial vasospasm, hypotension and hypovolaemia. Although the ^{133}Xe inhalational technique used in this study had limited spatial resolution, one further important observation was made about the distribution of CBF changes: unilateral carotid tree aneurysms were found to affect CBF in both hemispheres, although the reduction was greatest ipsilateral to the ruptured aneurysm. This observation is of great importance, and must be borne in mind when applying interhemisphere methods of comparison of rCBF in regional studies performed with SPECT (see below), in which relative rather than absolute CBF changes are calculated.

A more detailed assessment of the relationship between angiographic vasospasm and CBF was obtained by Voldby et al.[45] who performed CBF studies with intra-arterial ^{133}Xe, followed within 30 min by conventional cerebral angiography. These studies were carried out 3–13 days following subarachnoid haemorrhage, and were performed initially in close temporal relationship to diagnostic angiography, and subsequently when there was neurological deterioration. Additionally, a jugular venous catheter allowed calculation of the cerebral metabolic rate of oxygen from the arteriovenous oxygen difference. The authors were able to confirm the finding of the majority of previous reports of a strong association between vasospasm and regional cerebral blood flow, and in refuting the overall contribution of other changes such as hydrocephalus. Moreover, they found that the degree of arterial narrowing was important in determining CBF, with moderate narrowing (25–50%) not affecting CBF and severe narrowing (>50%) causing significant reductions. However, the limited resolution of the method prevented exact correlation between focal angiographic vasospasm and localized hypoperfusion.

Recent non-SPECT CBF studies

The technique of xenon-enhanced CT has provided useful supplementary data on the relationship between CBF, clinical symptoms, and ischaemic damage visible on structural scans. This method allows more localized and accurate estimation of blood flow than other xenon techniques, in particular the assessment of low flow areas. Yonas et al.[47] reported 51 patients, 14 of whom had neurological deterioration attributable to vasospasm, allowing first a comparison of CBF levels in symptomatic and asymptomatic cases, and second an indication of an ischaemic threshold by comparing CBF levels with the subsequent development of CT infarction, which developed in none of the 14 symptomatic cases. Flow values of < 15 ml/100 g/min were followed by delayed evidence of cerebral infarction, while values of > 18 ml/100 g/min could be symptomatic (altered conscious level and/or focal neurological signs) but were predominantly reversible and were not associated with CT infarction. Additional information was obtained from two of their symptomatic patients in whom hyperperfusion was found as an early sign of restored blood flow in CT-confirmed infarction, resembling CBF findings in patients with atherosclerotic ischaemic stroke.

PET has provided additional regional information about changes in cerebral flow and metabolism following subarachnoid haemorrhage. Available studies involve significantly fewer patients than other methods, however, and repeat measurements in individual patients are more limited than, for example, xenon CBF studies, so that extensive clinical and radiological correlations, and observation of the time course of changes cannot be made. Using the $^{15}O_2$ inhalation technique, Hino et al.[36] compared the findings in 13 patients with subarachnoid haemorrhage and 10 controls. The exact timing of PET was not stated, but the studies were performed 'during the period of delayed vasospasm in all patients', at which time angiographic vasospasm was present in nine of 11 subjects studied, and around which time nine patients had symptomatic vasospasm, consisting of focal neurological deficits in seven. Global metabolic suppression was demonstrated, confirming other reports[37], and cerebral blood volume was increased, probably as a result of compensatory dilatation of the parenchymal vessels in response to ischaemia. It was concluded that regional cerebral cortical hypoperfusion exists in patients with angiographic and/or clinical vasospasm, and that this exists against a global background of impaired supply and metabolism of oxygen. A combination of impaired metabolism of oxygen and impaired tissue supply of oxygen was supported by these findings. In a smaller study of four patients, two of whom had irreversible cerebral ischaemia, regional flow and oxygen metabolism were obviously lower in symptomatic sufferers[37]. Values were reduced bilaterally but were always lower within affected hemispheres, a finding which again illustrates the potential for error when interhemisphere comparison is made using SPECT.

While ischaemic thresholds are important, Powers et al.[37] have emphasized that the dynamic process of cerebral ischaemia and arterial vasospasm may affect interpretation of the results. In particular, the duration of hypoperfusion, the preceding metabolic status of the brain, and the susceptibility of different neurone populations to ischaemia will influence the final outcome of an ischaemic threat. This may explain why patients with subarachnoid haemorrhage have a higher ischaemic threshold than patients with acute ischaemia[36]. The failure of PET to identify consistent ischaemic thresholds for $CMRO_2$ has been illustrated in this and other studies, despite the fact that a reduction in $CMRO_2$ is one of the earliest changes after subarachnoid haemorrhage, occurring within the first 4 days, at which time CBF may be normal[36,37]. The problem of taking pictures which only relate to a short period of time during a changing physiological situation is illustrated by these studies. The measurement of cerebral blood volume can sometimes clarify haemodynamic relationships, but was not found to be useful by Powers et al., who comment that cerebral blood volume can be increased even in infarcted tissue[37]. Nevertheless, PET studies are clearly very valuable in that they parallel the results of regional perfusion studies using SPECT, and those of hemisphere CBF measurements using techniques such as inhaled xenon.

SPECT studies in subarachnoid haemorrhage

SPECT studies with HMPAO and [^{123}I]IMP in patients with subarachnoid haemorrhage have been reported since 1988. In a preliminary report of HMPAO SPECT, seven patients were studied with both angiography and MRI within 24 h of SPECT scanning. This early study showed that SPECT had a high sensitivity for lateralized and/or focal hypoperfusion abnormalities, which could be present in the absence of angiographic vasospasm[38]. The authors suggested that the location of hypoperfusion might help to define the source of bleeding in patients with bilateral or multiple aneurysms.

In a subsequent study using [^{123}I]IMP, 11 patients with aneurysmal subarachnoid haemorrhage were examined serially[43], comparing groups with reversible, irreversible, and fatal symptomatic vasospasm. The early images obtained following [^{123}I]IMP injection correspond to regional perfusion, but in this study a delayed image was also obtained 5 h after tracer injection. This was assessed as a possible indicator of reperfusion and cerebral metabolism[43]. Reversible ischaemic symptoms in five patients were accompanied by regions of moderate hypoperfusion and arterial narrowing. The CBF changes normalized with clinical recovery and CT remained normal. On the other hand, four patients with irreversible deficit showed deficits in [^{123}I]IMP uptake which were more dense in 5 h post-injection scans, and CT eventually showed infarction. In two patients who died with delayed vasospasm, earlier and more widespread changes were seen both on CBF and CT scans. The changes in the 5-h images were considered to be beneficial in distinguishing reversible from irreversible ischaemia, but the number of cases and data presented do not provide firm evidence for this conclusion.

More recent studies have examined the use of HMPAO SPECT to define the development of delayed ischaemic deficit, potentially the most useful clinical application for SPECT in patients with subarachnoid haemorrhage. CT and transcranial Doppler have been evaluated in this context, and will also be considered here. The extent and thickness of subarachnoid blood deposits on CT correlates with the risk of delayed vasospasm, but accurate prediction of neurological complications for individual cases has not proved possible. Further, patients presenting late, after subarachnoid blood becomes isodense with brain on CT, cannot be assessed in this way. The ability of transcranial Dopper to measure non-invasively increasing arterial velocities which accompany vasospasm allows prediction of delayed ischaemic deficit with moderate sensitivity and specificity[48], but does not allow examination of distal segments of the basal cerebral arteries, and anterior cerebral artery readings are difficult to interpret (compared with middle cerebral arteries) due to sharing of blood flow via the anterior communicating artery[40].

Correlations between CBF abnormalities and clinical grade, outcome and angiographic vasospasm, as established in earlier CBF studies (see above) are important, but more significant is the potential of matching the clinical ischaemic hemisphere with the site of regional perfusion abnormalities. Although some correlation was reported from PET[36] and ^{133}Xe-CT[47] studies, this experience was not uniform[37], and recent SPECT studies have provided more definitive information. In an initial study, Davies et al.[39] performed 18 HMPAO SPECT scans in 13 selected patients with subarachnoid haemorrhage, 10 of whom had aneurysms identified on cerebral angiography. SPECT was performed after the onset of neurological deficit in six of the 10 patients with positive angiograms, but was also available for one who was asymptomatic at the time of SPECT but subsequently developed CT-confirmed infarction. Focal hypoperfusion was found in all cases, confirming that SPECT is highly sensitive in detecting changes caused by symptomatic vasospasm. Patients without neurological deficit usually had normal SPECT scans, and patients with severe deficits had the greatest focal flow reductions, as assessed by inter-hemispheric comparison of signal counts.

In a further report, Davis et al.[40] selected 20 of 34 consecutive patients with subarachnoid haemorrhage for CBF studies. Results were compared between clinical groups without delayed ischaemia, with delayed ischaemia and lateralized neurological deficit, and with delayed ischaemia and non-lateralized deficit. Hypoperfusion matching the neurological deficit was detectable in nearly all patients, and a correlation with Doppler findings was evident, especially for those with lateralized deficit. Symptomatic patients with non-lateralized deficit tended to have anterior cerebral territory ischaemia, highlighting the contribution of this brain region to impaired general neurological status. Concordance between the two techniques was not complete in all patients, however, probably because SPECT shows blood flow at the microcirculatory level, while transcranial Doppler measures changes in the proximal cerebral arteries[38].

In a similar comparative study of transcranial Doppler and SPECT, we selected 20 patients for SPECT from a consecutive series of 121 cases[41]. This selection was made on the basis of a rapid rise (exceeding 50 cm/s in 24 h) to high levels of the velocity in the middle and/or the anterior cerebral arteries, as measured by transcranial Doppler. SPECT was repeated in six patients. The main aim was to perform SPECT scans prior to the onset of any neurological deficit, to assist in the prediction rather than the diagnosis of delayed ischaemic deficit. Fifteen patients were studied while asymptomatic: 14 showed abnormal perfusion patterns on SPECT, and 10 subsequently developed a focal neurological abnormality. Perfusion patterns correlated with the sites of increased Doppler velocities. In one patient without deficit the SPECT scan was normal. Four patients had zones of cerebral hypoperfusion but did not develop neurological deficit. These findings confirmed those of Davis et al.[39,40], although their studies were commonly performed after the onset of clinical deficit. We carried out repeat SPECT in three patients in whom the initial SPECT was performed before the onset of deficit; the perfusion defect increased in size in two of these cases.

The findings in follow-up CT scans were also of importance in relation to the SPECT changes in our study. The sites of SPECT perfusion defects correlated with the sites of infarction on CT in every case, although the former tended to involve larger areas, as has been shown in other studies[38,39].

Although we did not perform SPECT studies in patients who did not have a rapid rise in Doppler velocities, we were able to identify a perfusion defect on SPECT in all but one of our 15 cases examined prior to symptoms, and this corresponded anatomically in every case to the subsequent symptoms. The finding of increased transcranial Doppler velocities and perfusion defects in patients who remained asymptomatic is consistent with the reversible ischaemia which is more readily quantified by PET and xenon CT studies (see above). When correlation between SPECT, transcranial Doppler velocities and clinical findings was considered, the artery with the highest velocity usually corresponded to the site of the aneurysm and the site of perfusion defect on SPECT, although this correlation was clearly stronger for middle than anterior cerebral arteries. We concluded that transcranial Doppler velocities measurements obtained frequently in the early phase after subarachnoid haemorrhage help to identify patients at risk of delayed neurological deficit. Even in the group with false-positive results, in whom a rapid rise in transcranial Doppler velocities velocity occurs without subsequent neurological deficit, a zone of reduced CBF may be present which is insufficient to cause infarction. Further details of use of transcranial Doppler in aneurysmal subarachnoid haemorrhage are reviewed by Miller and Smith, in conjunction with CBF study results[49].

While the majority of studies have addressed the issue of vasospasm, CBF changes have also been described in association with acute hydrocephalus, a common and reversible complication of subarachnoid haemorrhage, which may cause delayed deterioration in conscious level. Hasan et al.[50] performed HMPAO SPECT in eight patients, four of whom had deterioration attributed to delayed ischaemia and four of whom showed hydrocephalus. The latter was diagnosed when the bicaudate index on initial or repeat CT scans showed a measurement above the 95th percentile of normal limits. In patients with delayed ischaemia, multiple and often bilateral areas of hypoperfusion were seen on SPECT, involving the posterior watershed areas in all patients and the anterior watershed areas in the majority. In patients with hydrocephalus, hypoperfusion was demonstrated in the region of the third and lateral ventricles, and in the basal temporal and frontal lobes. The areas of hypoperfusion were significantly larger than the size of ventricles on CT in all cases; by contrast, only limited flow defects were observed in brain cortex. Serial lumbar puncture was followed by improved cerebral perfusion in three of four patients with hydrocephalus, accompanied by clinical improvement; the final patient developed vasospasm while under study. Although the pattern of CBF disturbances in patients with delayed ischaemia differs from that in other SPECT studies[39–41], the observations relating to hydrocephalus are consistent with findings from patients with hydrocephalus of other causes. The changes are not explained by compression of cerebral tissue as a result of hydrocephalus, and the large size of the hypoperfused areas means that low flow measurements did not merely arise from inclusion of ventricle within the areas of sampling (the 'partial volume' effect). The exact explanation remains unclear, and may involve true hypoperfusion or hypometabolism in these areas. In conclusion, although CBF studies are not appropriate to diagnose hydrocephalus, the periventricular areas should be examined carefully on CBF studies and compared with CT scans when hydrocephalus is a potential cause of a patient's clinical deterioration. An additional feature seen in SPECT scans of patients with subarachnoid haemorrhage is a pattern of hyperperfusion, or of mixed areas of hypo- and hyperperfusion. Initial studies had insufficient resolution to enable identification of such appearances[38], and they have not been commented upon in some more recent HMPAO studies[38,40]. We found such mixed zones in four initial and one repeat SPECT scan in our 20 cases of subarachnoid haemorrhage selected for HMPAO SPECT. Three of the scans were performed before the onset of neurological deficit[41]. Similar patterns were found using [^{123}I]IMP during the phase of resolution of vaso-spasm[43]. Hyperperfusion has also been seen on Xe CT[51], as has luxury perfusion using PET scans[37].

Time course of CBF changes after subarachnoid haemorrhage

The exact time course of CBF changes varies according to the age and clinical condition of the patient, and some discrepancies have been obtained between studies employing different techniques. An initial fall in CBF (within the first week of subarachnoid haemorrhage) is invariable, and precedes the development of angiographically detectable vasospasm. Meyer et al.[46] observed a persistent decrease in cerebral blood flow throughout their 3-week study period, irrespective of the patient's clinical progress. On the contrary, Matsuda et al.[52] found that although some patients showed a progressive deterioration in the second week, there was a return to normal flow values by the second week in young patients with good outcome, and a recovery over the subsequent 3 weeks in most other patients. The initial reduction in CBF was no greater in patients of poor clinical grade than in other groups. These results are at variance with earlier findings, in which clinical grade was found to correlate with CBF levels[44–46]. Other inconsistencies have been observed, including rebound hyperaemia in one study[53], the finding of persistent low flow at up to one year in patients with neurological complications but good functional recovery in another[52], and normal results in the chronic period in a third[43]. These inconsistencies are likely to relate to the type of patients studied and differing methodologies used.

The relationship between cerebral blood flow, angiographic changes and clinical progress deserves further comment. Patients with delayed ischaemic deficit commonly have low CBF[35,47], especially in the second week[52]. One earlier study found no relationship between angiographic vasospasm and a regional reduction in CBF[35], but this association has been described by several other groups, and is particularly evident in cases with angiographic evidence of diffuse vasospasm[35,45,52]. The presence of angiographic narrowing does not necessarily indicate distal hypoperfusion, however[46], and cerebral blood flow depression may correlate more with clinical features than with angiographic vasospasm[52]. SPECT studies examining these relationships are, however, quite limited, as transcranial Doppler rather than angiography has been used to gauge arterial narrowing in the more recent studies[40,42].

Practical aspects of SPECT in subarachnoid haemorrhage

Isotope injection may be undertaken in the ward before transporting the patient to the CBF laboratory. Clearly, appropriate facilities are required for all patients, but patients of a poorer clinical grade require intensive nursing and anaesthetic support as well as portable monitoring equipment, to ensure safety throughout transfer and the 30 min scan time. The minimal head movement necessary during scan acquisition may be difficult to achieve in confused or dysphasic patients. Repeat SPECT studies, in particular to assess treatment effects, require additional expertise and patience.

Treatment effects and cerebrovascular reactivity in subarachnoid haemorrhage

Combined observations on the time course of cerebral blood flow obtained using SPECT and other CBF techniques, which suggest that global and regional hypoperfusion develops during the first week after bleeding, are consistent with the clinical finding of an increased risk of cerebral ischaemic complications when surgery is performed during the period of low CBF[46]. No prospective study to examine the validity of using CBF studies to guide the timing of surgery has been performed, but preoperative assessments of CBF do allow the patient's risk to be broadly defined[44]. The effects of surgery itself are also important. The normal increase in CBF after craniotomy which usually occurs in patients with subarachnoid haemorrhage is not seen in patients who have a subsequent poor outcome[44]. This observation is readily explained if surgery is considered as a prolonged ischaemic challenge.

The intraoperative autoregulatory CBF response to halothane-induced hypotension shows similar prognostic significance. Two distinct patterns of CBF response to this ischaemic challenge were shown in 20 patients studied by Pickard et al.[54]. In 15 of 20 patients, the CBF response to hypotension was a 58% increase; in the remaining five a fall of 48% was observed. Both groups contained patients with immediate postoperative deficits attributable to surgery, but the striking difference in CBF responses was closely matched by the late progress of the patients in the two groups: four of five patients with impaired CBF responses developed symptomatic delayed ischaemia, while only one of the 15 cases with a good response did so[54]. The increased CBF with hypotension in the good outcome group may represent a vasodilator effect of halothane; in patients with poor outcome, it is assumed that this response is impaired and/or the response to hypotension is impaired.

The effects of other interventions on CBF have been studied. Earlier studies with tranexamic acid showed varying results[46,52], but more recent studies provide useful data on the suitability and efficacy of current treatments. Hypertensive, hypervolaemic therapy is used to improve cerebral perfusion in patients with subarachnoid haemorrhage, but is usually reserved for patients who have undergone surgical aneurysm clipping, in view of the risk of preoperative rebleeding. This therapy has never been the subject of a controlled trial, but clinical studies have consistently demonstrated the reversal of focal neurological deficits after therapy is commenced, and frequent recurrence of the same symptoms and signs if an attempt is made to withdraw therapy at an early stage. Muizelaar[55] comments on the use of CBF studies to decide on use of hypertensive therapy, which is better avoided if patients have flow values well above the ischaemic threshold. Symon et al.[44] found that failure of the CBF response to volume expansion was a poor prognostic feature. Mendelow et al.[56] used CBF studies to assess this type of treatment, and in particular to guide therapy withdrawal. In eight patients, the administration of dopamine was associated with an increase in blood pressure and a return of abnormally low CBF readings towards normal. When dopamine was discontinued, blood pressure and CBF fell significantly; when patients deteriorated dopamine was restarted and CBF studies repeated during later withdrawal, so that 15 withdrawal procedures were monitored. Patients with a fall in CBF of >25% had a subsequent poor outcome compared with the remaining cases. It was provisionally concluded that such testing could guide withdrawal of hypertensive therapy. This type of reasoning could also be applied to other treatment strategies.

The assessment of therapeutic efficacy may in principle be assessed by paired measurements with SPECT, potentially using a split dose technique. One advantage of HMPAO SPECT in this setting is that therapy can be given as soon as HMPAO is injected, and need not be delayed until after the baseline scan is performed. The use of CBF studies to assess the benefits of cerebral angioplasty have also recently been reported, and complement transcranial Doppler velocities recordings from the appropriate vessels[42].

Arteriovenous malformations

A high flow arteriovenous malformation (AVM) in the brain causes local and remote cerebral blood flow abnormalities. Shunting of blood directly from the arterial to the venous circulation leads to local cerebral hypoperfusion[57–59], although the extent and severity of intracerebral steal is not readily predictable from angiographic appearances[57]. Secondary diaschisis involving large portions of the ipsilateral cerebral hemisphere, and typically also the contralateral cerebellar hemisphere, have also been shown by CBF studies, both preoperatively and intraoperatively[58]. In all of 37 patients studied preoperatively, at least one area of both ipsilateral and contralateral hypoperfusion was identified, although the size and severity of the areas of abnormal perfusion varied considerably[59]. This situation clearly resembles the diaschisis of ischaemic stroke (see above). Prolonged diaschisis has been suggested as the cause of structural MRI changes in the thalamus and brain stem of such patients[58]: the degree of damage depends on the size and position of the AVM and is particularly severe for those involving the basal ganglia. Neuropathological studies confirm this phenomenon[58]. These CBF observations help to explain why clinical deterioration develops in the absence of direct complications from the AVM itself, such as bleeding.

Studies of the reactivity of CBF in cerebral tissue surrounding an AVM have provided conflicting results. Tarr et al.[59], using xenon CT and intravenous acetazolamide, found that vasoreactivity was normal before embolization of AVMs and diminished thereafter, particularly at 6–10 days. This change was sometimes associated with symptomatic deterioration. Contrary results were found by another group, whose initial observation of a patient who developed fatal hyperaemia postoperatively, and who had an abnormal increase in acetazolamide responsiveness, led to a more detailed study of 35 cases: Batjer et al.[57] found enhanced vasodilatation in areas adjacent to the AVM using ^{133}Xe SPECT. Moreover, there was an association between this abnormality enhanced vasoreactivity and the subsequent development of hyperaemic complications; this enhancement was significant for ipsilateral areas of abnormal flow and showed a similar trend in contralateral regions, as well as for regions of maximal hypoperfusion. An additional correlation between CBF and angiographic findings was made: patients with angiographic steal (reduced contrast flow to vessels adjacent to an AVM) had the greatest depression of regional CBF.

The importance of defining the reactivity differences of vessels adjacent to an AVM relates to the rare but potentially fatal complication of hyperaemia leading to cerebral oedema and/or intracerebral haemorrhage, which may occur following AVM removal. The primary cause of this problem is likely to be the abrupt reshunting of high flow into the vasculature surrounding the obliterated AVM. The risk of 'breakthrough' damage clearly depends on the integrity of the reactive arterial mechanisms in this area, and the current contradictory findings require further evaluation, to develop pharmacological techniques of limiting postoperative damage, and to guide the nature and timing of operative intervention.

Conclusions

The ready availability of SPECT makes it an exciting facility for the evaluation of patients with disorders of the cerebral vasculature. For many years measurement of cerebral blood flow was regarded as an important parameter for assessment, and yet the limitations of the techniques available never allowed it to find its way into routine clinical practice. Indeed in later years standard text books on cerebrovascular disease have indicated a diminishing interest and importance. The development of easy-to-use tracers of cerebral blood flow, and the ease with which images can be obtained using standard instrumentation such as a rotating gamma camera have now made regional blood flow studies available to most clinicians. HMPAO SPECT is currently a non-quantifiable technique, and it cannot be assumed, particularly in the context of cerebrovascular disease, that blood flow and metabolism are always coupled. Nonetheless the evaluation of changes in brain perfusion provided by SPECT should now guide clinicians to more logical use of present treatments, as well as allowing the assessment of new treatments. Hanson et al.[60] have recently reviewed the value of SPECT in acute stroke therapeutic trials. They suggest that a measurement of change in cerebral perfusion is a valid assessment of hemisphere stroke severity in the hyperacute setting, and should be useful for the selection and stratification of patients entering clinical therapeutic trials. For the future, the development of radioligands that allow an assessment of the role of receptors and their neurotransmitters in the ischaemic process is to be welcomed. The full exploitation of such techniques will depend on the specificity of the ligands, and the availability of SPECT images of a sufficient degree of spatial resolution.

Acknowledgements

The authors are grateful for the support of the Chest Heart and Stroke Association (Scotland), the Robertson Trust, and the Institute of Neurological Sciences Research Trust.

References

1. Reivich M, Ginsberg M, Slater R, et al. Alterations in regional cerebral hemodynamics and metabolism produced by focal cerebral ischemia. Eur Neurol. 1978;17(Suppl. 1):9–16.
2. Ono S, Yanagimoto S, Mimura H, et al. [Usefulness of 99mTc-ECD SPECT in diseases of the central nerve system: special reference to a comparison with 123I-IMP and 99mTc-HM-PAO SPECT]. Kaku Igaku. 1992;29:1429–40.
3. Itoh H, Iida H, Murakami M, et al.[A method for measurement of regional cerebral blood flow using N-isopropyl-p[123I]iodoamphetamine (123I-IMP) SPECT: two scans with one point blood sampling technique]. Kaku Igaku. 1992;29:1193–200.
4. Skinhoj E, Hoedt Rasmussen K, Paulson OB, Lassen NA. Regional cerebral blood flow and its autoregulation in patients with transient focal cerebral ischemic attacks. Neurology. 1970;20:485–93.
5. Bogousslavsky J, Delaloye Bischof A, Regli F, Delaloye B. Prolonged hypoperfusion and early stroke after transient ischemic attack. Stroke. 1990;21:40–6.

6. Lassen NA, Olsen TS, Hogaard K. Incomplete infarction: a CT negative irreversible ischaemic brain lesion. J Cereb Blood Flow Metab. 1983;3(Suppl. 1):602–3.
7. Chollet F, Celsis P, Clanet M, et al. SPECT study of cerebral blood flow reactivity after acetazolamide in patients with transient ischemic attacks. Stroke. 1989;20:458–64.
8. Bullock R, Mendelow AD, Bone I, Patterson J, Macleod WN, Allardice G. Cerebral blood flow and CO2 responsiveness as an indicator of collateral reserve capacity in patients with carotid arterial disease. Br J Surg. 1985;72:348–51.
9. Sugiyama Y, Haba K, Yokoyama K. [SPECT evaluation of effect of cerebral vasodilator by the subtraction method using Tc-99m HMPAO]. No To Shinkei. 1993;45:139–42.
10. Pacia SV, Boss A, Fayad P, Smith EO, Hoffer PB, Brass LM. Single photon emission computed tomography (SPECT) in distal field hypoperfusion. J Neuroimag. 1991;1:31–3.
11. Raynaud C, Rancurel G, Samson Y, et al. Pathophysiologic study of chronic infarcts with I-123 isopropyl iodo-amphetamine (IMP): the importance of periinfarct area. Stroke. 1987;18:21–9.
12. Raynaud C, Rancurel G, Tzourio N, et al. SPECT analysis of recent cerebral infarction. Stroke. 1989;20:192–204.
13. De Roo M, Mortelmans L, Devos P, et al. Clinical experience with Tc-99m HM-PAO high resolution SPECT of the brain in patients with cerebrovascular accidents. Eur J Nucl Med. 1989;15:9–15.
14. Cordes M, Henkes H, Roll D, et al. Subacute and chronic cerebral infarctions: SPECT and gadolinium-DTPA enhanced MR imaging. J Comput Assist Tomogr. 1989;13:567–71.
15. Baird AE, Donnan GA. Prognostic value of reperfusion during first 48 hours of ischaemic stroke. Lancet. 1993;342:236.
16. Lee RG, Hill TC, Holman BL, Royal HD, O'Leary DH, Clouse ME. Predictive value of perfusion defect size using N-isopropyl-(I-123)-p-iodoamphetamine emission tomography in acute stroke. J Neurosurg. 1984;61:449–52.
17. Bushnell DL, Gupta S, Mlcoch AG, Barnes WE. Prediction of language and neurologic recovery after cerebral infarction with SPECT imaging using N-isopropyl-p-(I 123) iodoamphetamine. Arch Neurol. 1989;46:665–9.
18. Giubilei F, Lenzi GL, Di Piero V, et al. Predictive value of brain perfusion single-photon emission computed tomography in acute ischemic stroke. Stroke. 1990;21:895–900.
19. Baird AE, Donnan GA, Austin M, Newton MR, McKay WJ. Preliminary experience with 99mTc-HMPAO SPECT in cerebral ischaemia. Clin Exp Neurol. 1991;28:43–9.
20. Rossell A, Landini G, Castagnali A, Vannuchi L. Nimodipine therapy of acute focal ischaemia: a clinical study with an assessment of regional cerebral blood flow by SPECT. Rec Prog Med. 1992;83:85–8.
21. Kempinsky WH. Experimental study of distant effects of acute focal brain injury. Arch Neurol Psych. 1958;376:389.
22. Baron JC, Bousser MG, Comar D, Duquesnoy N. Crossed cerebellar diaschisis: a remote function depression secondary to supratentorial infarction in man. J Cerebr Blood Flow Metab. 1981;(Suppl):500–1.
23. Bowler JV, Costa DC, Jones BE, Steiner TJ, Wade JP. High resolution SPECT, small deep infarcts and diaschisis. J R Soc Med. 1992;85:142–6.
24. Bogousslavsky J, Miklossy J, Regli F, Deruaz JP, Assal G, Delaloye B. Subcortical neglect: neuropsychological, SPECT, and neuropathological correlations with anterior choroidal artery territory infarction. Ann Neurol. 1988;23:448–52.
25. Maurer AH, Siegel JA, Comerota AJ, Morgan NA, Johnson MH. SPECT quantification of cerebral ischaemia before and after carotid endarterectomy. J Nucl Med. 1990;31:1412–20.
26. Hayashida K, Nishimura T, Uehara T, Kaminaga T, Hirose Y. Raise up 99mTc-HMPAO brain SPECT for detecting the changes in cerebral perfusion pressure. Nippon Igaku Hoshasen Gakkai Zasshi. 1992;52:1036–8.
27. Algotsson L, Ryding E, Rehncrona S, Messeter K. Cerebral blood flow during carotid endarterectomy determined by three dimensional SPECT measurement; relation to preoperative risk assessment. Eur J Vasc Surg. 1993;7:46–53.
28. Di Piero V, Lenzi GL, Collice M, et al. Long-term noninvasive single photon emission

computed tomography monitoring of perfusional changes after EC-IC bypass surgery. J Neurol Neurosurg Psychiatry. 1987;50:988–96.
29. Batjer HH, Devous MDS, Purdy PD, Mickey B, Bonte FJ, Samson D. Improvement in regional cerebral blood flow and cerebral vasoreactivity after extracranial-intracranial arterial bypass. Neurosurgery. 1988;22:913–9.
30. Spierings ELH. Arteriovenous shunting in migraine. In: Amerg WK, Van Neuten WK, Mauquier A, editors. The pharmacological basis of migraine therapy. London: Pitman, 1984:36–49.
31. Andersen AR, Friberg L, Olsen TS, Olesen J. Delayed hyperemia following hypoperfusion in classic migraine. Single photon emission computed tomographic demonstration. Arch Neurol. 1988;45:154–9.
32. Schlake HP, Bottger IG, Grotemeyer KH, et al. Single photon emission computed tomography with technetium-99m hexamethyl propylenamino oxime in the pain-free interval of migraine and cluster headache. Eur Neurol. 1990;30:153–6.
33. Stillhard G, Landis T, Schiess R, Regard M, Sialer G. Bitemporal hypoperfusion in transient global amnesia: 99mTc-HM-PAO SPECT and neuropsychological findings during and after an attack. J Neurol Neurosurg Psychiatry. 1990;53:339–42.
34. Cohen MB, Graham LS, Lake R, et al. Diagnosis of Alzheimer's disease and multiple infarct dementia by tomographic imaging of iodine-123 IMP. J Nucl Med. 1986;27:769–74.
35. Mickey B, Vorstrup S, Voldby B, Lindewald H, Harmsen A, Lassen NA. Serial measurement of regional cerebral blood flow in patients with SAH using 133Xe inhalation and emission computed tomography. J Neurosurg. 1984;60:916–22.
36. Hino A, Mizukawa N, Tenjin H, et al. Postoperative hemodynamic and metabolic changes in patients with subarachnoid hemorrhage. Stroke. 1989;20:1504–10.
37. Powers WJ, Grubb RL, Baker RP, Mintin MA, Raichle ME. Regional cerebral blood flow and metabolism in reversible ischaemia due to vasospasm. Determination by positron emission tomography. J Neurosurg. 1985;62:539–46.
38. Rawluk D, Smith FW, Deans HE, Gemmell HG, MacDonald AF. Technetium 99m HMPAO scanning in patients with subarachnoid haemorrhage: a preliminary study. Br J Radiol. 1988; 61:26–9.
39. Davis S, Andrews J, Lichtenstein M, et al. A single-photon emission computed tomography study of hypoperfusion after subarachnoid haemorrhage. Stroke. 1990;21:252–9.
40. Davis SM, Andrews JT, Lichtenstein M, Rossiter SC, Kaye AH, Hopper J. Correlations between cerebral arterial velocities, blood flow, and delayed ischemia after subarachnoid hemorrhage. Stroke. 1992;23:492–7.
41. Grosset DG, Straiton J, duTrevou M, Bullock R. Prediction of symptomatic vasospasm after subarachnoid hemorrhage by rapidly increasing transcranial Doppler velocity and cerebral blood flow changes. Stroke. 1992;23:674–9.
42. Lewis DH, Eskridge JM, Newell DW, et al. Brain SPECT and the effect of cerebral angioplasty in delayed ischemia due to vasospasm. J Nucl Med. 1992;33:1789–96.
43. Nakagawara JP, Nakamura J, Matsuzaki T, et al. Assessment of hemodynamic changes due to delayed cerebral vasospasm in subarachnoid hemorrhage using SPECT and 123 I-IMP. In: Wilkins RH, editor. Cerebral vasospasm. New York: Raven Press, 1988:79–85.
44. Symon L, Compton J, Redmond S, Rosenstein J, Momma F. Bedside monitoring in subarachnoid hemorrhage: evoked responses, hemispherical blood flow, and flow velocity measurements. In: Wilkins RH, editor. Cerebral vasospasm. New York: Raven Press, 1988: 73–8.
45. Voldby B, Enevoldsen EM, Jensen FT. Regional CBF, intraventricular pressure, and cerebral metabolism in patients with ruptured intracranial aneurysms. J Neurosurg. 1985;62:48–58.
46. Meyer CHA, Lowe D, Meyer M, Richardson PL, Neil-Dwyer G. Progressive change in cerebral blood flow during the first three weeks after subarachnoid haemorrhage. Neurosurg. 1983;12:58–76.
47. Yonas H, Sekhar L, Johnson D, Gur D. Determination of irreversible ischemia by xenon-enhanced computed tomographic monitoring of cerebral blood flow in patients with symptomatic vasospasm. Neurosurg. 1989;24:368–72.

48. Grosset DG, Straiton J, McDonald I, Cockburn M, Bullock R. Predictive value of transcranial Doppler sonography for the development of delayed ischaemic deficit after subarachnoid haemorrhage. J Neurosurg. 1993;78:183–7.
49. Miller JD, Smith RR. Transcranial Doppler sonography in aneurysmal subarachnoid haemorrhage. Cerebrovasc Brain Metab Rev. 1994;6:31–46.
50. Hasan D, vanPeski J, Loeve I, Krennig EP, Vermeulen M. Single photon emission computed tomography in patients with acute hydrocephalus or with cerebral ischaemia after subarachnoid haemorrhage. J Neurol Neurosurg Psych. 1991;54:490–3.
51. Bucheit F, Boyer P. Review of treatment of symptomatic vasospasm with nimodipine. Acta Neurochirur. 1988;45(Suppl.):51–5.
52. Matsuda M, Shiino A, Handa J. Sequential changes of cerebral blood flow after aneurysmal subarachnoid haemorrhage. Acta Neurochir Wien. 1990;105:98–106.
53. Tsementzis SA, Meyer CH, Hitchcock ER. Cerebral blood flow in patients with a subarachnoid haemorrhage during treatment with tranexamic acid. Neurochirurgia Stuttg. 1992;35:74–8.
54. Pickard JD, Matheson M, Patterson J, Wyper D. Prediction of late ischemic complications after cerebral aneurysm surgery by the intraoperative measurement of cerebral blood flow. J Neurosurg. 1980;53:305–8.
55. Muizelaar JP. Cerebral blood flow measurements in the diagnosis and treatment of aneurysmal 'vasospasm'. In: Wilkins RH, editor. Cerebral vasospasm. New York: Raven Press, 1988:63–72.
56. Mendelow AD, Dharker S, Patterson J, Nath F, Teasdale GM. The dopamine withdrawal test following surgery for intracranial aneurysms. J Neurol Neurosurg Psych. 1986;49:35–8.
57. Batjer HH, Devous MDS. The use of acetazolamide-enhanced regional cerebral blood flow measurement to predict risk to arteriovenous malformation patients. Neurosurgery. 1992;31: 213–7.
58. Tanaka K, Yonekawa Y, Kaku Y, Kazekawa K. Arteriovenous malformation and diaschisis. Acta Neurochir Wien. 1993;120:26–32.
59. Tarr RW, Johnson DW, Horton JA, et al. Impaired cerebral vasoreactivity after embolization of arteriovenous malformations: assessment with serial acetazolamide challenge xenon CT. Am J Neuroradiol. 1991;12:417–23.
60. Hanson SK, Grotta JC, Rhoades H, et al. Value of single-photon emission-computed tomography in acute stroke therapeutic trials. Stroke. 1993;24:1322–9.

5. SPECT in dementia, schizophrenia and other psychiatric disorders

MARTIN TURNER and DAVID J. WYPER

Introduction

The psychiatric disorders dealt with in this chapter represent a major public health problem. The prevalence of dementia is 10% in the over 65 age group, and 20% in the over 80s. Half of these cases are due to Alzheimer's disease (AD)[1]. Major depression has a lifetime prevalence of 4.9–8.7% in women and 2.3–4.4% in men, mania has a lifetime prevalence of 1.2%, and schizophrenia of 1%[2–4]. The pathophysiology of these disorders is only partly understood. Alzheimer's disease poses a particular problem, as treatments are beginning to become available and their evaluation is hampered by discrepancies between clinical and pathological diagnostic criteria.

Functional neuroimaging is contributing significantly to the understanding of the pathological processes underlying the above diseases. The development of SPECT, PET and latterly MRI techniques which permit the examination of the functional activity of the brain in different psychological and pathological states is enabling the study of different disease processes during the life of the patient, where previously such knowledge could only be gained indirectly through the study of pathological samples obtained postmortem. The potential rewards for psychiatry which may arise from the development of these capabilities include a progression from disease classification based on descriptive observation of behaviour to a classification based on more directly quantifiable measure of changes in brain activity and functioning.

Current methods available for the examination of brain function enable cerebral activity to be assessed through markers of cerebral blood flow, with the assumption that these measures reflect the metabolic demands for oxygen and, therefore, local brain activity. More recently, the development of radiolabelled ligands which bind to neurotransmitter receptors has allowed the measurement of receptor densities and the level of receptor occupancy by neurotransmitters within the living brain. These techniques are particularly pertinent given the central role of neurotransmitter hypotheses in current models of the development of psychiatric disease.

The appropriate development of these techniques requires that appropriate questions are asked, and that suitable tests are applied. This consideration is nowhere more pertinent than in psychiatry, where symptomatology reflects subtle dynamic functional change rather than the structural changes seen in other disease processes. This subtlety is further complicated by moment-to-moment

R. Duncan (ed): SPECT Imaging of the Brain, 131–160.

and day-to-day changes in mental state in those with psychiatric illness; the chronic schizophrenic may still suffer mood swings and episodes of anxiety and agitation. Such changes can have profound effects on the localized use of glucose and local cerebral blood flow. It is readily apparent that failure to control for these variables will make some functional brain imaging techniques vulnerable to immense intra-individual fluctuation. Fortunately, the development of many neuropsychological tests over the years has provided a rich source of material which allows for discrete activation of different regions of the brain by the performance of particular tasks. These 'neuropsychological probes'[5] may then be used to study patient groups which can be standardized for cognitive tasks at the time of testing. The comparison of patients with a particular psychiatric disease process with matched normal control groups performing the same tasks permit evaluation of impairment of objective task performance with the functional activity of the brain. A body of normative data describing the effects of execution of somatosensory tasks on cerebral blood flow and glucose utilization exists from PET research. These studies have identified a wide range of variables which may influence cerebral blood flow, including the significance of handedness as well as the type of task performed[6,7]. General effects, such as the influence of anxiety and the consequent physical sequelae such as hyperventilation may also be studied and the impact of such emotional states on the functional brain scan gauged (SPECT techniques have been shown to be generally less stressful than PET studies, lower levels of autonomic arousal resulting from the test procedure[8]). The study of regional cerebral blood flow may also be sensitive to the effects of ageing. Gur[9] studied the effect of age on cerebral blood flow in resting and activation states; advanced age was associated with reductions in overall blood flow with a particularly pronounced reduction in anterior regions. The extent and pattern of rCBF during activation conditions was not affected by age.

The influence of pharmacotherapy upon brain function must also be considered, both in evaluating the reports and papers published by researchers and in considering the results of individual SPECT scans requested for clinical reasons. Both prescribed and illicit drugs influence brain function, the specific effects depending on the dose and type of drugs involved. Bartlett et al.[10] demonstrated markedly different effects following treatment of chronic schizophrenics with either thiothixene or haloperidol: the former increased whole brain glucose utilization while the latter decreased glucose utilization. The use of hallucinogenic drugs by volunteers has also shown changes in cerebral blood flow patterns with a predominantly right-sided hyperfrontal pattern which correlates strongly with the degree of induced psychotic psychopathology[11].

Research into functional changes which may occur in brain metabolism during psychiatric illness has a well established pedigree, with early work being published by Ingvar and Franzen in 1974[12]. These early studies employed the technique of unilateral (left hemisphere) intra-arterial injection of ^{133}Xe to measure regional cerebral blood flow (see Chapter 1) in populations of chronic schizophrenic patients. The results of these studies showed low frontal and high posterior cerebral flows in older patients relative to a control group of alcoholics. The test included activated and resting states. Such reports were largely ignored for many years until the emergence of newer technologies which have permitted easier, less invasive approaches to the study of brain metabolic activity.

At the macroscopic level psychiatric disorders are generally 'functional' rather than 'structural' in nature and, potentially, more insight can be gained from PET and SPECT imaging than from CT or conventional MRI. For example, a study comparing [^{99}Tc^{m}]HMPAO SPECT abnormalities with CT patterns in patients with clinically diagnosed Alzheimer's disease who subsequently had postmortem confirmation of the disease and also in elderly control subjects, showed that SPECT and CT scanning both disclosed a higher incidence of abnormality in AD patients than in controls. The difference between patients and controls was about 50% greater with SPECT than with CT[13]. Around 50% of the SPECT deficits occurred in brain regions which were normal on CT, showing that atrophy was not the sole cause of these deficits.

Most research activity has so far been concentrated on schizophrenia and dementia, although there is a steady flow of papers addressing other issues in psychiatry, including the affective disorders, neurotic conditions such as obsessive-compulsive disorder and substance abuse.

Dementia

Alzheimer's disease

AD is the most common form of dementia, accounting for at least half of all cases of dementia examined at necropsy. It is a primary disorder of nerve cells, characterized by neuronal death and extensive brain atrophy. Although it is over 80 years since Alzheimer presented his case report, the disease that bears his name has assumed major importance only in the last decade. In recent years the application of biochemical and molecular genetic approaches to AD has increased understanding at a cellular level. Along with these advances in basic science, there have also been advances in the clinical understanding of AD. The importance of proper clinical evaluation and investigation of patients presenting with signs of dementia is now widely recognized. Definitive diagnosis, however, requires histological examination, which is usually only possible postmortem as brain biopsy is usually difficult to justify clinically. More accurate premortem diagnosis allows better evaluation of management approaches and service planning, as well as providing a better base for research into pathogenesis.

The neuropathological changes in AD are characteristic: neuronal loss, senile plaques and neurofibrillary tangles. It has therefore been argued that AD is a single disease entity, with its own causes and natural course. Such a conclusion gained support from the discovery of the cholinergic deficit in AD. As central cholinergic systems had already been implicated in pharmacological models of memory, it seemed plausible that the memory deficits of AD could be caused by selective loss of cholinergic neurons[14]. However other neurochemical deficits also characterize AD, and the disease process is more complex than previously thought. It is not known whether the different patterns of neurochemical deficit represent different stages of a single disease process or different processes. The cause of neuronal death in AD is not known, but it does appear to follow a consistent pattern, leading to a characteristic clinical syndrome. It is possible that AD is the endpoint of different causal pathways, each leading to this pattern of neuronal damage.

Clinical features, course and outcome

The onset of AD, although difficult to determine with any precision, is usually after the age of 40 years. Insidious memory impairment, dysphasia, dyspraxia and visuospatial abnormalities, and the progressive development of global dementia are usually recognized as its cardinal features; behavioural, affective, psychotic and neurological symptoms may also occur[15]. Memory difficulties occur as the earliest feature more often in AD than in any other type of presenile dementia[16]. Along with intellectual and personality deterioration, focal symptoms appear. Language dysfunction is common, with the development of expressive and later, receptive, dysphasia. Parietal lobe dysfunction leads to dyspraxia and agnosia, although neuronal loss and tangle formation are often less severe in parietal regions than in frontal and temporal areas. However, parietal association cortex has major connections with superior and inferior temporal cortex and with prefrontal cortex[17] and structural and neurochemical deficits in frontal cortex may, therefore, contribute to dyspraxia[18]. Late in the disease process neurological disabilities develop, with increased tone, myoclonus, Parkinsonian features and extensor plantar reflexes.

Before SPECT imaging became routinely available a variety of criteria had been proposed in order to improve the clinical diagnosis of AD, including evidence of cerebral atrophy on X-ray CT with progression documented by serial observation. Neuropathological confirmation is important for the audit of clinical diagnosis and remains essential before a diagnosis of definite AD can be made.

Relationship between early and late forms of Alzheimer's disease

Historically, dementia occurring after the age of 65 years has been considered separately from that occurring in younger patients. The pathological changes in early onset AD appear to be more severe than in SDAT (senile dementia of Alzheimer type). Detailed comparisons between the pathology of younger and older patients show that reduction in brain weight is greater in presenile cases than in SDAT[19]. Further support for the subdivision comes from the differences in neocortical noradrenaline concentrations, choline acetyltransferase activity, plaques and tangles. Similarly, neurotransmitter deficits in postmortem brain are more pronounced and more widespread in younger patients with an earlier onset dying before the age of 70–80 years, than in older, late-onset patients dying after the age of 70–80 years[20]. The neurochemical deficits of earlier onset AD tend to be predominantly in fronto-temporal cortex, while in older cases are more confined to temporal cortex[20].

Mapping patterns of functional abnormality using SPECT imaging

Studies using ^{133}Xe and [^{123}I]IMP: Using the ^{133}Xe inhalation SPECT method, 19 patients with probable AD were studied by Bonte et al.[21]. Compared with age- and sex-matched control subjects, AD patients showed reduced rCBF in bilateral temporal regions, and in some this was more marked on the left side. Left frontal, and bilateral parietal and occipital reductions in flow were also detected in more severely affected patients. The utility of [^{123}I]IMP for SPECT in dementia was assessed in a small study of five patients with probable AD: diffuse symmetrical reductions in relative [^{123}I]IMP uptake, particularly in parieto-occipital areas, were demonstrated[22]. In a larger study in Aberdeen[23] which included 14 senile patients with probable AD, SPECT scanning was used to determine relative [^{123}I]IMP uptake in 10 symmetrical regions. All of the patients with probable AD showed bilateral reduction of uptake in temporo-parietal-occipital regions, and in many patients posterior frontal flow deficits were also detected. Interestingly, in 64% of the AD group, MRI was normal in posterior regions, reinforcing the observation that SPECT abnormalities are not always evident on structural imaging.

Using standardized diagnostic criteria, 15 patients with probable AD were compared with a group of nine age-matched controls using [^{123}I]IMP and a rotating gamma camera SPECT system[24]. The patient group consisted of mixed presenile and senile probable AD types, with onset of illness after 65 years in eight patients and before 65 years in the remainder. Marked reductions in cortical/cerebellum ratios of tracer uptake occurred in all cortical areas of the AD group compared with controls, but were most marked (20–30% decrease) in posterior parietal, posterior temporal and frontal areas. Overall, no asymmetries were detected, but marked side to side differences did occur in individual cases. In a different study[25], nine patients with probable AD were compared with five non-matched control subjects using [^{123}I]IMP SPECT, including neuropsychological testing. In all patients with probable AD there was marked reduction in relative [^{123}I]IMP uptake in bilateral temporo-parietal cortex, and this was significantly associated with severity of dementia, as measured by the Mini-Mental State Examination score. Rather ambitiously, on the basis of this study, Jagust et al.[25] concluded that the [^{123}I]IMP SPECT method was able to distinguish AD from control subjects, even in very early stages of the disease. The authors also emphasized the importance of impaired temporo-parietal rCBF and speculated about its significance in the pathogenesis of AD, and possible utility for staging the progression of the disease.

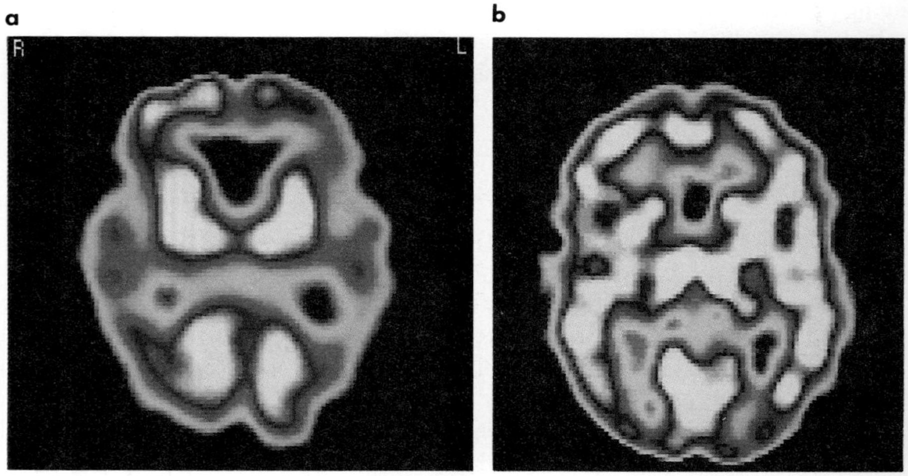

Figure 5.1. SPECT scans of a patient with clinically diagnosed Alzheimer's disease (A) compared with an age-matched control (B). The patient was an 85-year-old lady who was admitted to a long stay bed around the time of this investigation. She had a CAMCOG score of zero. Alzheimer's disease was confirmed postmortem. Note the low uptake of [^{99}Tc^{m}]HMPAO in posterior temporal regions compared with the control. There is also reduced uptake in the left lateral frontal region.

HMPAO: For routine imaging ^{99}Tc^{m} is more widely available than ^{123}I and is much less expensive. There is high correlation between relative rCBF, measured using [^{99}Tc^{m}]HMPAO SPECT, and corresponding absolute rCBF values, measured with ^{133}Xe inhalation, in both normal subjects and probable AD patients[26]. Furthermore, in dementia and other neuropsychiatric disorders, [^{99}Tc^{m}]HMPAO SPECT has an advantage in that the tracer is administered intravenously and has a distribution in brain that is relatively stable; allowing sufficient time for imaging and patient co-operation.

Diagnostic utility of SPECT in the differential diagnosis of dementia

There are quite distinct patterns of distribution of perfusion deficit in the various forms of dementia (Figures 5.1, 5.2). The most common pattern in AD is one of symmetrically reduced uptake in the posterior temporal/parietal cortex, although reduced frontal uptake has also been reported in patients with subsequent postmortem confirmation of the disease. In Korsakoff's psychosis the perfusion pattern is more generalized and impaired flow in frontal regions correlates with impaired performance on tests of memory and orientation[27]. Vascular dementia can present in a variety of ways. The deficit is often unilateral and easily distinguished from AD. The pattern in multi-infarct dementia (MID) is of two or more areas of reduced uptake and is usually asymmetrical and seldom involves parietal cortex. Differential diagnosis of depression from dementia is important

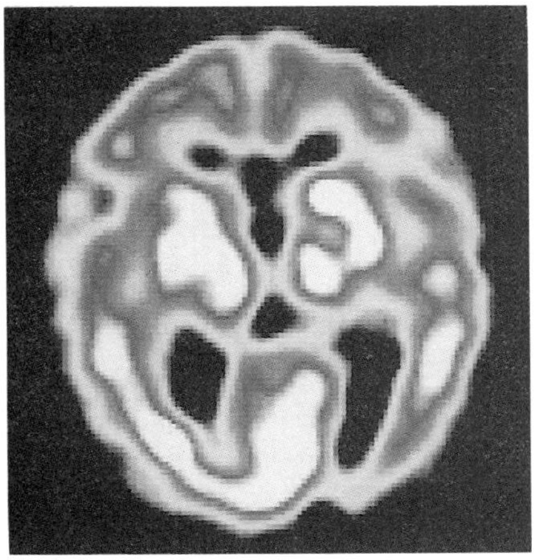

Figure 5.2. SPECT scans using [$^{99}Tc^{m}$] HMPAO of an 87-year-old man admitted unconscious with focal neurological signs of right extensor plantar and primitive reflexes. His CT showed low density changes in white matter in both hemispheres consistent with ischaemic damage and widening of CSF spaces. The SPECT appearance is very different from the Alzheimer pattern with several regions of markedly reduced tracer uptake, in particular in left and right temporal and left occipital regions and is consistent with multi-infarct dementia.

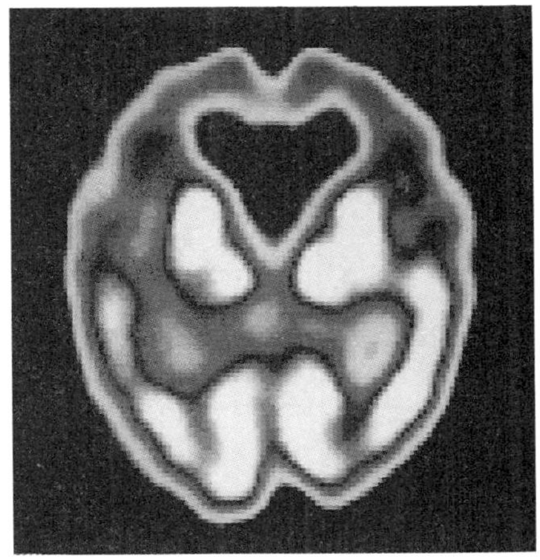

Figure 5.3. SPECT scan of an 88-year-old man presenting with loss of interest, apathy and avolition. He was prone to mood swings, although he was always courteous. His CAMCOG score was 38, with poor performance on attention, memory, language and abstraction and normal perception and praxis. He did not deteriorate significantly over a 14-month period. At postmortem he showed none of the characteristics of Alzheimer's disease, with no plaques. There was no evidence of Pick or Lewy bodies. The postmortem diagnosis was frontal lobe dementia.

as effective treatments are available: the perfusion pattern in depression can be either normal or show reduced frontal uptake, while in Huntington's disease there is decreased uptake in the caudate nucleus, with normal parietal uptake.

Frontal lobe dementia and Pick's disease are characterized by marked bilateral anterior flow deficits (Figure 5.3). In a qualitative study using [$^{99}Tc^{m}$]HMPAO, Neary et al.[28] examined 21 patients with presumed AD and nine others with frontal lobe dementia. The tomograms were evaluated by inspection, and classified according to whether they showed anterior or posterior areas of reduced uptake. In the AD group, 33% had a posterior deficit, a similar percentage a mixed anterior and posterior deficit, and the remainder had either reduced anterior uptake or no detectable abnormality. In two-thirds of the AD patients, visuo-spatial impairment was associated with reduced posterior flow. In patients classified as having 'frontal dementia' on the basis of clinical features, 80% had, not unexpectedly, reduced frontal flow. In a comparable qualitative study of 17 AD patients, and 10 MID patients, Gemmell et al.[29] used $^{99}Tc^{m}$ HMPAO SPECT to investigate whether the two groups could be distinguished by pattern of uptake. Tomograms were divided into 10 regions and their degree of abnormality

Table 5.1 Probability of disease by scintigraphic pattern.

Perfusion pattern	Probability of Alzheimer's disease (%)	Probability of other diseases (%)
A (normal)	19	81 (including NL)
B (bilateral T/P)	82	18 (18 PDD)
C (bilateral T/P + other)	77	23 (12 PDD) (12 VD)
D (unilateral T/P)	57	43 (22 VD) (14 PDD) (7 PPA)
E (frontal)	42	58 (43 PSP) (15 HIV)
F (other)	18	82 (many diseases)
G (multiple small)	0	100 (HIV)

NL = normal; PDD = Parkinson's disease with dementia; VD = vascular dementia; PPA = primary progressive aphasia; HIV = HIV dementia; PSP = progressive supranuclear palsy.
Taken from Holman et al.[31] with permission

scored on a scale of 1–3. All patients also had magnetic resonance imaging, and no focal abnormalities were detected in the AD patients. Reduced [^{99}Tc^{m}]HMPAO uptake was more common in the AD group than the MID group, especially in bilateral temporo-parietal-occipital regions. Both studies suggested that inspection of SPECT images may permit better diagnosis of dementia type and that AD patients tended to have bilateral posterior deficits. Reduced flow in bilateral temporal and posterior parietal regions was also found in older AD patients with aphasia and apraxia, and associations were found between memory loss and temporal lobe activity, and between language and left sided tracer uptake[30].

In an important prospective study of the diagnostic value of SPECT in dementia, Holman et al. found that a probability for AD of 82% in patients with posterior temporal SPECT deficits presenting with memory loss at a tertiary referral centre[31]. They classified [^{99}Tc^{m}]HMPAO SPECT patterns into 7 categories:

A: normal;
B: bilateral posterior temporal and/or parietal cortex defects;
C: bilateral posterior temporal and/or parietal cortex defects with additional defects;
D: unilateral posterior temporal and/or parietal cortex defects with or without additional defects;
E: frontal cortex defects only;
F: other large (>7 cm) defects;
G: multiple small (<7 cm) cortical defects.

They investigated the diagnostic value of this classification in a prospective study in which disease was determined at clinical follow-up averaging 10.1 months. The distribution of disease by scintigraphic pattern is shown in Table 5.1 and the probability of disease by scintigraphic pattern is shown in Table 5.2. This

Table 5.2. Distribution of disease by scintigraphic pattern.

SPECT pattern	Total patients per pattern	Final diagnosis AD	PDD	VD	PSP	PD	HIV	TGA	Trauma	SLE	PPA	EP	DP	NL
A	21	4	2	3	0	0	2	1	0	1	0	2	3	3
B	17	14	3	0	0	0	0	0	0	0	0	0	0	0
C	26	20	3	3	0	0	0	0	0	0	0	0	0	0
D	14	8	2	3	0	0	0	0	0	0	1	0	0	0
E	7	3	0	0	3	0	0	0	0	1	0	0	0	0
F	17	3	1	2	1	2	1	4	2	0	0	1	0	0
G	11	0	0	0	0	0	11	0	0	0	0	0	0	0
Total patients per diagnosis	113	52	11	11	4	2	14	5	2	2	1	3	3	3

AD = Alzheimer's disease; PDD = Parkinson's disease dementia; VD = vascular dementia; PSP = progressive supranuclear palsy; PD = Pick's disease; HIV = HIV dementia; TGA = transient global amnesia; SLE = systemic lupus erythematosus; PPA = primary progressive aphasia; DP = depression; NL = no CNS disease. Taken from Holman et al.[31] with permission study suggests that the predictive value of bilateral temporoparietal defects for Alzheimer's disease is high. Although this was an extremely valuable study and provided a good indication of diagnostic efficacy within a relatively short period of time, it lacked the definitive end-point of autopsy confirmation of the disease.

The Optima project (Oxford Project to Investigate Memory and Ageing) is addressing the diagnostic value of neuroimaging in a study which includes histopathology[32]. Studies of the most accepted clinical criteria – the Diagnostic and Statistical Manual of Mental Disorders, Third Edition, Revised (DSMIII-R) and the National Institute of Neurological and Communicative Disorders and Stroke and the Alzheimer's Disease and Related Disorders Association Work Group (NINCDS-ADRDA) – suggest that their diagnostic accuracy varies considerably, but is around 75% on average. To date Jobst and colleagues in the OPTIMA study have performed SPECT and CT in 71 cases, 47 with AD, 16 with other dementias and eight controls, along with 84 living controls. The CT was undertaken in a 'temporal lobe' orientation similar to the plane used for SPECT imaging in epilepsy. They measured the narrowest thickness of the medial temporal lobe, on either the right or left side, at the level of the brain stem between its anterior and posterior margins. SPECT was first assessed quantitatively by z-score analysis against controls. This numerical data was used as a guide for a final consensus report to categorize scans using a graded 4-point scale; 0 for normal tracer uptake; 1 for a mild, 2 moderate and 3 severe uptake deficit. A score of 2 or 3 in the parieto-temporal cortex was regarded as significant. The authors have examined the sensitivity (percentage of patients with positive histopathology and a diagnosis of AD) and specificity (percentage of patients with negative histopathology and diagnosed as not having AD)[33] of the diagnostic imaging procedures. Specificity is perhaps the more important parameter in the case of AD. The findings to date are that SPECT alone has a 96% sensitivity and 89% specificity for AD; temporal lobe oriented CT alone has a 94% sensitivity and 93% specificity, and both procedures combined have a sensitivity of 90% and specificity of 97%. In other words the diagnostic false-positive rate of both tests combined is only 3%. The 84 living controls have been included in these computations. These figures are surprisingly good by comparison with other reports. It may be that the Oxford study is focusing on a selected group of patients relative to those seen at other routine referral centres, and certainly the patients reported to date have been within a few years of death. Nonetheless, this is the largest study with pathology as its diagnostic end-point, and it does demonstrate the value of combining information from more than one investigation. The results of this study are also of importance in providing evidence of the association between medial temporal atrophy and parieto-temporal hypoperfusion, suggesting that the latter may result from loss or damage to pyramidal cells in medial temporal lobes.

Table 5.3. Positive associations between cognitive performances and rCBF.

ROI	CAMCOG	Memory	Language	Praxis
Rf	*		*	*
Lf	*		*	*
Rt	*		*	*
Lt	**		**	*
Rpt	**	*	**	**
Lpt	**	*	**	**
Rhf				*
Lhf				*
Rp	**	*	*	**
Lp	**	*	**	**

*Correlations significant at an individual 5% level.
**Correlations significant at an overall 5% level when applying a Bonferroni correction.

Effects in frontal cortex in Alzheimer's disease

In the early stages of AD, during which the patient goes from normal to showing mild/moderate signs of dementia, almost every functional imaging study has shown that the posterior association cortex is the most likely to be affected. However, Haxby et al. using PET to investigate regional cerebral metabolic rate for glucose in patients with mild/moderate AD[34] and a longitudinal follow-up of 20 patients showed that parietal/frontal metabolic ratios and their correlated neuropsychological patterns were stable over time, as dementia worsened, indicating that in patients with moderate to severe AD metabolic reductions in the premotor cortex are as severe as those in the parietal association cortex.

In a recent study completed by the authors' group, investigating the association between regional functional decline measured using SPECT with [^{99}Tc^{m}] HMPAO and cognitive decline measured by the MMSE, the frontal regions were those most significantly associated with decline in MMSE. This new finding, along with the data of Haxby et al., suggests that the frontal functional decline which becomes evident as the disease progresses also affects the cognitive decline of the patient.

SPECT deficits and cognitive performance

The pattern of perfusion deficit in AD has been shown to correlate with abnormalities in cognitive performance[35]. In a SPECT study of 26 older AD patients, there was reduced rCBF in cortical regions, particularly bilaterally in posterior temporal areas but also in left temporal, left parietal and bilateral frontal regions, with relative sparing of occipital regions. Numerous associations between cognitive performance and rCBF were found (Table 5.3), which might reflect the severity of disease in this patient group, although the non-specificity of the cognitive battery used may have contributed. The fact that no associations between memory and rCBF were detected is not surprising, in view of the pronounced floor effects that occur in AD with most simple tests of memory. The

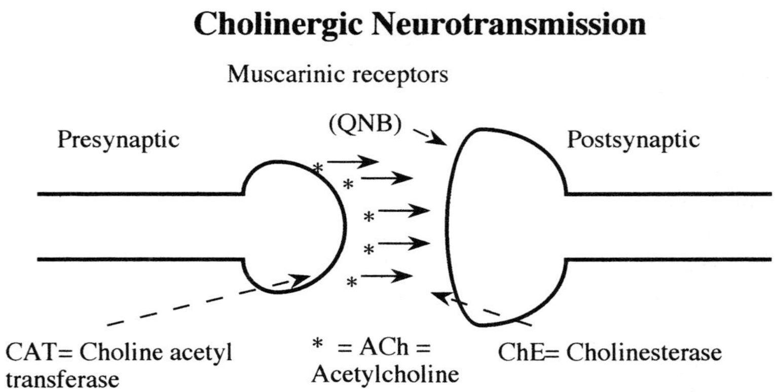

Figure 5.4. A schematic illustration of the muscarinic acetylcholine neuroreceptor. By inhibiting the action of the enzyme cholinesterase the synaptic concentration of acetylcholine can be increased. Quinuclidinyl benzilate (QNB) is an antagonist which competes for binding with the endogenous neurotransmitter acetylcholine. When labelled with ^{123}I it can be used as a tracer of muscarinic receptor density.

correspondence between neuropsychological deficits and SPECT in AD was investigated by Goldenberg et al.[36], and the results interpreted in terms of two dimensions of organization, along which rCBF measures and cognitive test scores co-varied in a systematic fashion. In a fronto-temporo-parietal axis, naming, constructional praxis and visual recognition memory were associated with parieto-temporal regions, and semantic memory and behaviour more associated with frontal areas. In a second left-right axis of organization, the ordering of test results was partly consistent with what is known about the lateralization of neuropsychological deficits from studies of localized lesions (e.g. with language-based tasks on the left).

SPECT and therapeutic intervention

The cholinergic hypothesis: Figure 5.4 illustrates the muscarinic acetylcholine neuroreceptor in schematic form. There is considerable evidence, both from postmortem data and from biopsy specimens, of acetylcholine depletion in patients with AD. A possible therapeutic strategy is, therefore, to increase the concentration of acetylcholine by reducing the action of acetylcholinesterase, the enzyme which breaks it down in the synapse. For this approach to work effectively it is important that postsynaptic receptors, both muscarinic and nicotinic, are intact. Investigations are described below which demonstrate the use of SPECT in mapping muscarinic receptor density and then in evaluating alterations in regional functional activity produced by cholinesterase inhibitors.

Are muscarinic receptors affected in AD? Various therapeutic agents have been developed with the aim of improving acetylcholinergic neurotransmission, but clinical studies have shown variable results, with cognitive improvement measured in some, but not all, patients. Postmortem studies have shown normal, increased or decreased muscarinic receptor density. As described in Chapter 1, SPECT can be used along with radiolabelled neuroreceptor ligands to measure neuroreceptor density. Muscarinic receptor density can be imaged in vivo using [^{123}I]QNB SPECT. Detailed characterization of the interaction of 4-iodo-QNB with receptor sites in homogenates of the corpus striatum in vitro has demonstrated that this ligand binds almost exclusively to a receptor population which is sensitive to atropine at doses highly selective for muscarinic receptors. The non-atropine sensitive portion (approximately 1%) represents non-specific binding and binding to other receptors. There is no evidence from in vitro studies that binding to receptors other than muscarinic contributes to the specific binding observed. The synthesis of (R,R)[^{123}I]QNB has been described by Owens et al.[37]. Radio-iodination is achieved using a Cu(I) assisted nucleophilic exchange reaction. The precursor for labelling, (R,R)127I-QNB is first fully characterized by standard analytical techniques: ^{1}H and ^{13}C NMR and UV, IR and mass spectrometry. Analysis of the labelled product is performed on an RPC18 HPLC column. The material has a radiochemical purity of 96% and a specific activity of 400 Ci/mmol. A dose of 160 MBq of the radioligand is administered to each subject. Scanning can be undertaken at various intervals after injection but the optimum period for scanning is at approximately 24 h. By this time the non-specific washout has cleared, but there is still sufficient activity in the source to enable good quality scans to be obtained with a high quality neuroSPECT imager as ^{123}I has a physical half-life of 13.2 h. The brain uptake of [^{123}I]QNB is slow, reaching a peak after about 20 h. If the uptake was rapid the distribution would be governed by delivery (i.e. blood flow) as well as by receptor density, but the slow uptake approximates to an equilibrium model and virtually eliminates the blood flow component in the tracer distribution. This enables a fairly pure measurement of receptor density to be obtained. One uncertainty, however, is the relative affinity of [^{123}I]QNB for the various subtypes of muscarinic receptor. Although it is thought to have a greater affinity for the M1 subtype followed by M2, more investigation of this is required.

Weinberger et al.[38] studied 14 patients with dementia (12 with AD) and 11 healthy controls using [^{123}I]QNB. Reduced IQNB uptake in temporal and posterior/temporal regions significantly separated AD patients from controls but they comment on the problems of interpreting these findings caused by the uncertainty about the degree to which brain atrophy and cell loss contribute to the focal reductions in uptake of IQNB. To further elucidate this the same group compared IQNB SPECT to fluorodeoxyglucose (FDG) PET in seven patients with dementia (five with AD) and five normal volunteers[39]. They found a dissociation, with [^{123}I]QNB SPECT scans showing larger deficits than FDG PET scans, but in one patient the [^{123}I]QNB scan was normal despite focal deficits on FDG PET.

This work was extended by Wyper et al.[40], who compared [^{123}I]QNB uptake with the pattern of regional cerebral blood flow measured using [^{99}Tc^{m}]HMPAO,

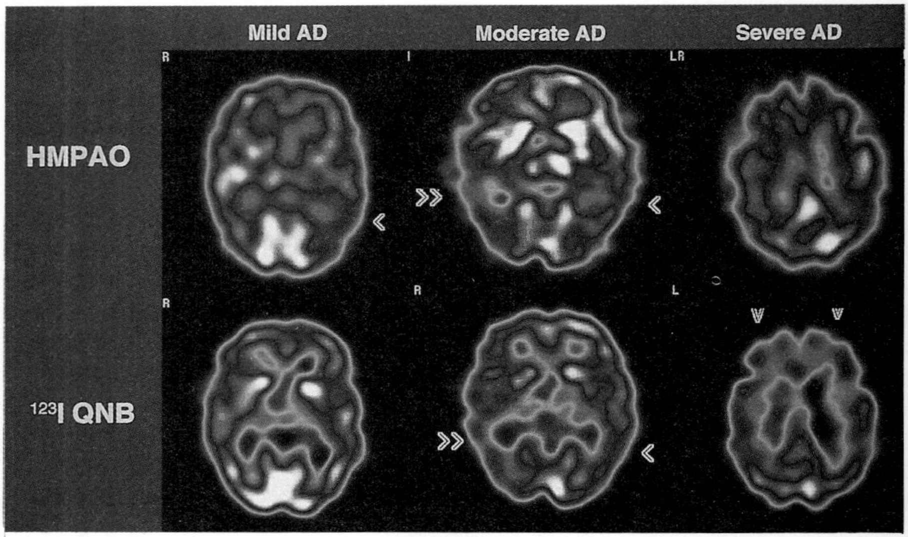

Figure 5.5. SPECT scans of the distribution of [^{123}I]QNB at 21 h compared with [^{99}Tc^{m}]HMPAO in patients with clinically diagnosed Alzheimer's disease and varying severity of cognitive impairment. The top scans are [^{99}Tc^{m}]HMPAO and the lower ones [^{123}I]QNB. The patients are: (a) a mildly demented AD patient with CAMCOG of 62; (b) a moderately demented patient with CAMCOG of 32; (c) a severely demented patients who was given a CAMCOG score of zero as she was untestable. Relative to [^{99}Tc^{m}]HMPAO (which gives a measure of general regional function as determined by blood flow), the focal deficits in the [^{123}I]QNB pattern at 21 h (which is governed by muscarinic receptor density) appear to be less marked in the mild case and more marked in the severe case. This suggests that muscarinic receptors are not selectively affected until late in the disease process. (For colour version of this Figure see page 184.)

which was used to give a measure of non-specific regional function. The results of this study are illustrated in Figure 5.5 (see also colour section) which shows the [^{123}I]QNB and [^{99}Tc^{m}]HMPAO distribution patterns in patients with mild, moderate and severe AD. The patient with mild AD showed no evidence of a reduction in muscarinic receptor density despite a slight generalized functional deficit apparent on the HMPAO scan. The patient with moderate dementia and a CAMCOG of 32 showed a reduction in receptor density in the posterior temporal/inferior parietal region comparable in magnitude to the general functional HMPAO deficit. Only the severely demented patients, who could not perform any of the cognitive tasks, displayed regional deficits in muscarinic receptor density which were in excess of the generalized HMPAO deficits. This study was performed using only eight patients and four controls, and the results need further corroboration. It does, however, demonstrate the power of SPECT to investigate specific alterations in neurochemistry in the disordered brain.

Points to consider in radioligand studies

In the above study, interpretation of [^{123}I]QNB images alone was difficult due to uncertainty about the effects of cell loss. The analytical approach adopted was to compare deficits in [^{123}I]QNB (as determined by deviation from regional uptake in control subjects) with deficits in regional cerebral blood flow measured using [^{99}Tc^{m}]HMPAO. The energies of the main gamma emissions from ^{123}I (160 keV) and ^{99}Tc^{m} (140 keV) are similar, allowing imaging with similar scanner characteristics.

Three patterns were observed in patients with AD when deficits in [^{123}I]QNB binding were compared to deficits in ^{99}Tc^{m} HMPAO uptake; first, deficits which are broadly similar (in magnitude and location) in both sets of scans; second, deficits which are more marked in HMPAO uptake than with [^{123}I]QNB as tracers; third, the deficits in which [^{123}I]QNB binding are more marked than those with the flow tracer. To assist in interpretation of these findings consideration should be given to the fundamental mechanisms which determine the alterations in blood flow distribution (imaged with [^{99}Tc^{m}]HMPAO) and muscarinic receptors (imaged with [^{123}I]QNB). Approximately 50% of cortical energy production (and of cortical blood flow continuously delivering the substances for energy production) is due to synaptic transmission. Local reductions in blood flow in AD may reflect reductions in neuronal activity in these areas; such an effect would not necessarily be associated with alterations in muscarinic binding sites. In addition, those changes in AD which indirectly reduce synaptic transmission, for example loss of cortical neurones or reductions in the number of dendrites which will reduce the number of synapses, are also likely to be major components of reductions in cortical blood flow in AD. Whether the reduction in blood flow due to cell loss (which may not be apparent on CT) is accompanied by a loss of IQNB binding sites will depend upon which neuronal elements have degenerated in particular AD patients and whether muscarinic receptors are located on those elements. Even if the post-synaptic acetylcholine receptors are not selectively affected in AD, the IQNB uptake measured by SPECT could still be reduced as a result of this cell loss. Therefore the receptor density per unit volume of cortical tissue could be normal, but the density of [^{123}I]QNB binding measured by SPECT will be reduced. In most cases the [^{123}I]QNB deficits, possibly due to cell loss alone, were not as marked as the [^{99}Tc^{m}]HMPAO deficits, which would be affected by atrophy, loss of synapses and probably any other factors which could disrupt synaptic transmission or neuronal function.

In vivo results compared with in vitro findings

The number of in vitro studies reporting a reduction in muscarinic receptors is balanced by an approximately equal number showing no change, while a smaller number report an increase. There have been suggestions that these discrepancies may be due to variations in the contribution of receptor subtypes to the measurements in the different assays, but even when individual receptor subtypes are examined the evidence for change is equivocal. Probst et al.[41] examined alterations in the number of muscarinic receptor density in relation to a quantitative measurement of CA1 pyramidal cells and found that the ratio of muscarinic receptors to pyramidal cells was increased in AD. They suggest that this is consistent with the hypothesis that compensatory increases in muscarinic receptors are possible in intact neurones, but that beyond a certain threshold of neuronal degeneration this response is no longer possible. The measurements reported in this in vivo study provide evidence supporting this observation. Disproportionately large deficits in [^{123}I]QNB binding were not found in six of the patients and were observed only in the two most severe cases, both of whom were untestable on the cognitive battery.

The weight of evidence from all sources suggests, therefore, that cholinergic receptors are intact in mild/moderate AD, and it remains possible that cholinesterase inhibition or other methods of boosting acetylcholine neurotransmission could be of value in improving cognitive performance.

Assessment of pharmaceutical intervention

It is recognized that evaluation of putative cognitive enhancers is often difficult, and increasing use is being made of functional imaging methods to examine alterations in brain metabolism or perfusion induced by pharmacological intervention. Several studies have been undertaken of the effect of acetylcholinesterase inhibitors on regional cerebral blood flow. Geaney et al.[42] reported significant increases in relative rCBF only in bilateral posterior parietal and temporal regions. In contrast, Hunter et al.[43], normalizing their results to white matter, found an increase in uptake in frontal regions of interest with relatively greater increase on the left side of the cortex. Neither of these studies reported the relationship between local blood flow changes and cognitive impairment after physostigmine. However, Ebmeier et al.[44] reported the effects of a single oral dose of velnacrine maleate on word and object recognition memory and regional uptake of [^{99}Tc^{m}]HMPAO using SPECT in patients with AD. In this study word recognition memory was marginally improved 2 h after administration of 75 mg of velnacrine. With the same dose of velnacrine, a relative increase in superior frontal uptake of [^{99}Tc^{m}]HMPAO was shown using SPECT. These results were interpreted as indicating increased regional perfusion and metabolism as a consequence of cholinergic stimulation. Furthermore, the more cognitively impaired patients appeared to show a greater increase in tracer uptake after velnacrine, suggesting cholinergic hypersensitivity in the brains of AD patients.

Hunter et al.[45] extended this work by investigating the effects of chronic velnacrine treatment. This study took place in the context of a preliminary phase of a placebo-controlled investigation of the effects of velnacrine on cognition. This preliminary phase consisted of AD patients receiving a 3-week treatment with velnacrine in order to determine who were possible ‘responders’ and ‘non-

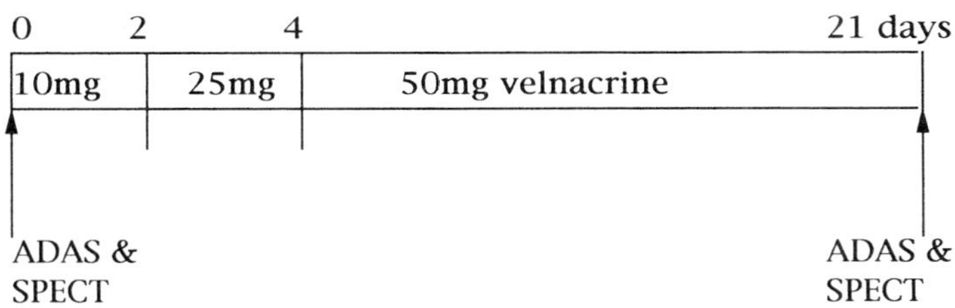

Figure 5.6. The experimental paradigm showing the timing of SPECT and cognitive testing in the assessment of the action of velnacrine.

responders'. Patients who responded to this preliminary phase were then entered into the drug trial proper. This afforded the opportunity to investigate rCBF changes occurring over 3 weeks of treatment with velnacrine in 'responder' patients and 'non-responder' patients with AD. The hypothesis under test was that velnacrine would cause relative increases in cerebral blood flow in AD patients whose cognition improved with velnacrine but not in those patients whose cognition appeared not to improve with velnacrine treatment.

Ten patients suffering from probable AD and meeting DSM III-R criteria for dementia[46] and NINCDS-ADRDA criteria for probable AD[47] were selected. Eight female and two male patients with a mean age of 69.6 (SD 6.9) years were studied. All patients had MMSE scores in the range of 10–24, mean 16.2 (SD 3.8) and mean illness duration of 34.8 (SD 19.6) months. The experimental design employed is shown in Figure 5.6. The AD assessment scale (ADAS) measurement and SPECT were carried out immediately before patients started drug treatment. Treatment with velnacrine gradually increased to a maximum dose of 50 mg three times per day. At the end of 21 days' treatment, SPECT and ADAS were repeated. Parallel forms of ADAS were used. 'Responders' to velnacrine were defined as those with an ADAS decrease of >4; all others were defined as 'non-responders'. Software allowing three-dimensional construction of brain images was used in order to ensure that identical slices were compared in the pre- and post-treatment SPECT images. Data from regions of interest (ROIs) were normalized using the average uptake in all ROIs and converted into z-scores, as described in Chapter 1, by comparison with a normal control data set ($n=11$) made available from other studies. Those ROIs with a z-score (see Chapter 1) of less than -2.0 were included in the analysis of uptake pre- and post-velnacrine, thereby focusing on those regions of the brain most affected by the disease. Although data were normalized to the average uptake in all regions, the normalization method was not critical as it has little effect on the measurement of change pre- and post-velnacrine. The results demonstrated a positive association between reduction in rCBF deficit and improvement in ADAS score as shown in the relationship between mean change in rCBF in all 'affected' brain regions and change in ADAS score pre- and post-velnacrine (Figure 5.7).

This study demonstrated the use of SPECT to study the chronic effects of a cholinesterase inhibitor on regional cerebral blood flow in AD patients in rela-

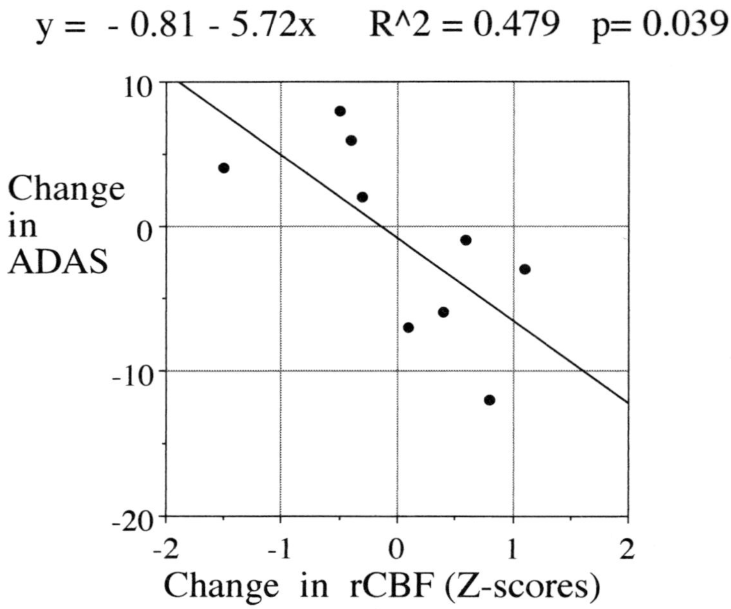

Figure 5.7. The results of corresponding changes in ADAS compared with changes in regional uptake of the CBF tracer [^{99}Tc^{m}]HMPAO in the most abnormal brain perfusion regions. Note that a high ADAS score means a poorer performance.

tion to changes in cognition and showed a good agreement between change in relative rCBF in affected brain regions and ADAS change after velnacrine. The inability of SPECT to provide absolute measurements of cerebral blood flow constitutes a limitation, as global changes in perfusion cannot be detected. The method used in this study was designed to focus on the regions of the brain most affected by the disease and to study changes in these regions relative to the brain as a whole. A relative improvement of perfusion in the most affected region is equivalent to the restoration of a more normal perfusion pattern. The results described above, although preliminary, suggest that this method of analysis is promising and that SPECT imaging has the potential to provide a physiological measurement to underpin cognitive testing in the evaluation of therapeutic agents.

Other neurotransmitter alterations in AD

It would be surprising if the neurodegeneration in AD was restricted to the cholinergic system. There is growing evidence for both GABA and glutamate involvement.

GABA is the predominant inhibitory neurotransmitter of the cerebral cortex and the hippocampus. It acts on two classes of receptors GABAa and GABAb. GABA is allosterically modulated by benzodiazepines, and the ‘benzodiazepine receptor' is a binding site on the GABAa receptor. There is growing evidence that the GABA receptor plays an important role in the pathology of AD[48]. Postmortem studies using radiolabelled ligands have shown a reduction in GABAa binding in the cortex and the hippocampus. Postmortem studies also show GABAergic nerve endings may degenerate in AD. A study using SPECT with [^{99}Tc^{m}]HMPAO and with the ligand ^{123}I-labelled iomazenil demonstrated normal blood flow images in the frontal cortex of AD patients in the presence of diminished benzodiazepine receptor concentration. This suggests that receptor changes may precede blood flow changes and may have implications in early diagnosis of the disease[49]. This loss of GABA activity in brain may contribute to the clinical signs of fear and anxiety in patients with AD.

It is known that antagonism of cortical glutamate receptors, such as the NMDA subtype impairs memory, the core symptom in AD. Although the evidence for reductions in the neurotransmitter glutamate is not conclusive, it is possible that increasing NMDA receptor activity may be of therapeutic benefit in AD. Excessive stimulation of the NMDA receptor is an important cause of cerebral cell death in some kinds of acute brain damage, such as stroke or head injury. It has been proposed that this may be an important mechanism in the destruction of neurones in AD.

A partial agonist of the NMDA glutamate receptor (cycloserine) is currently being tested in clinical trials. It has been suggested that if glutamate receptor activation is reduced this drug may act as an agonist, improving memory. If there is over-stimulation of the glutamate receptor, which may cause neurodegeneration, it is hoped that the drug will act as an antagonist. It is therefore important in the development of treatments of AD to determine whether glutamate-mediated NMDA receptor activity is reduced or increased. This is currently being investigated by SPECT studies using the glutamate antagonist MK801 labelled with ^{123}I. If this ligand is shown to be effective it can be used to investigate whether cycloserine is an antagonist or an agonist: i.e. if it increases or decreases NMDA receptor activation in AD patients.

Schizophrenia

The study of schizophrenia has long been at the heart of psychiatric research, and the new techniques available have provided a new dimension to the understanding of this disease process. In 1983, Sheppard[50] published a study of 12 acute episode schizophrenics. Using the newly developed technique of PET, the study compared mostly drug-free schizophrenics with a control group of volunteers matched for age, sex and handedness. The test was performed at rest, and demonstrated no reduction of frontal blood flow. A difference did exist between controls and patients, controls showing asymmetry on laterality while the patients showed no difference between hemispheres. The advent of SPECT allowed the development of relatively inexpensive techniques to examine functional brain activity, and this has vastly increased the number of imaging studies. In a seminal work in 1986, Weinberger reported on a study group of 20 chronic schizophrenic patients entered into a study using ^{133}Xe inhalation in a variety of test conditions. The patient group showed differences from the control group which were dependent on the challenge condition. At rest the patients had a significantly reduced relative blood flow to the dorsolateral prefrontal cortex area (DLPFC). The activation tasks comprised a number sorting test and the Wisconsin Card Sort Test (WCS). During number matching no specific region distinguished patients from controls. During the WCS, both rCBF absolute and rCBF relative to DLPFC significantly distinguished patients from controls. Controls showed a clear increase in DLPFC rCBF, while patients did not. These changes were specific to DLPFC, and rCBF to this area correlated positively with WCS performance. The patients involved were all drug free for a period of at least 4 weeks prior to the study.

The studies so far undertaken have at times yielded contradictory findings, but this may reflect the stage of the illness at which patients have been studied and the rather 'mixed' groups which have often been placed together in single studies. The majority of studies of schizophrenia have employed regional cerebral blood flow measurements, with an inference that these factors are necessarily linked to brain activity. Thus the assumption is that regional differences in blood flow which exist between schizophrenic patients and identified control subjects can be attributed to the disease process. The dynamic nature of SPECT, which measures these regions of blood flow at the time of radiolabelling, is most sensitive to the mental tasks and state of arousal at the time of the labelling. The studies which are now being reported in the literature often attempt to standardize the likely intellectual activity of the individual so that comparisons can be made between individuals and for the same individual investigated at different stages in the disease process. Standardization may be achieved through ensuring a resting state (unstimulated), and this may be achieved through the investigation being performed in a dimly lit room, with low levels of background noise, often with the subject having their eyes closed at the time of labelling. Such an approach cannot control for the internal thought processes of the subject; this is often likely to include a high level of anxiety, sometimes with hallucinations and frequently with delusional ideas. These factors may all influence the regions of the brain which are activated. In order to standardize to some extent, the use of activated challenge conditions has become more commonplace, with the most frequently used techniques being those of the Wisconsin Card Sort Test (WCS) or simple verbal fluency tests. These have the advantage of standardizing the likely intellectual activity of subjects at the time of radiolabelling, although it is far from clear whether the different test paradigms permit direct comparison between different study populations.

The increasing use of these tests has been partly driven by earlier findings which have indicated abnormalities of blood flow in the frontal region, particularly the dorsolateral prefrontal cortex. These tests are based on previously developed neuropsychometric tests which involve activation of these areas of the brain. Results using these challenge conditions have to some extent replicated

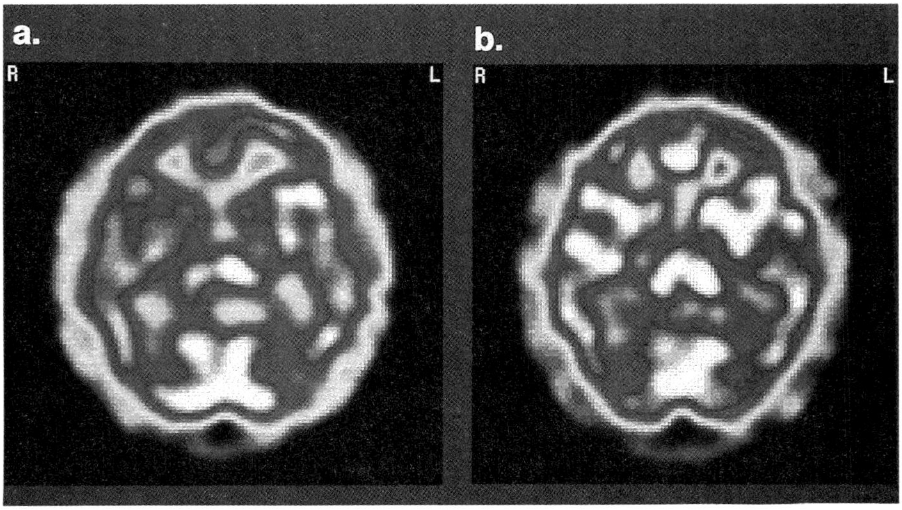

Figure 5.8. SPECT scans with [^{99}Tc^{m}]HMPAO at the level of the basal ganglia from (a) a drug-naive schizophrenic patient and (b) the same patient on neuroleptics, showing an increase in blood flow in the basal ganglia. (For colour version of this figure see page 185.)

earlier findings, the predominant finding being that of hypofrontality. This finding is far from universal, but careful assessment of subgroups of schizophrenics based upon the presence of positive or negative symptoms and the presence or absence of neuroleptics gives a clearer picture. Overall hypofrontality correlates with the severity of negative symptoms of schizophrenia, an effect which appears to be largely independent of neuroleptic drugs[51]. In a more acute population with positive symptoms, resting blood flows were increased in the left hemisphere, correlating with the degree of positive psychotic symptoms[52]. On activation tests using the WCS, reduced left frontal rCBF correlated with poor task performance. The failure of rCBF to DLPFC to increase in schizophrenia during performance of the WCS appears characteristic of the disease state[53]. Similar findings have been demonstrated using word-fluency activation[54] and neuroleptic-related effects are also seen in treated schizophrenics. The most consistent of these findings are increases in the rCBF in the caudate nucleus and thalamus (Figure 5.8; see also colour section). The subject shown in Figure 5.8 is a 23-year-old male presenting with a first episode of psychotic illness which satisfied the DSM-III-R criteria for schizophrenia. The subject was recruited into the Scottish Schizophrenia Research Group first episode schizophrenia study. Entry criteria into the study also required subjects to have never been exposed to any neuroleptic medication, to have no history of substance abuse and to have no concomitant physical illness. Subjects satisfying these criteria receive their first SPECT scan within 3 days of admission to hospital, and a second scan after 6 months of treatment with neuroleptics, during which they are monitored closely using a battery of psychometric tests and symptom rating scales. At the time of the initial scan, this subject scored 34 on the Brief Psychiatric Rating Scale

(BPRS); at the follow up scan this had changed to 13. The SPECT scans were performed following injection of [^{99}Tc^{m}]HMPAO while the subject performed a verbal fluency task involving the naming of words beginning with a particular letter. This continued over a 6-min period, with the letter given to the patient changing after each successive minute. Injection took place at the start of minute three. Throughout this time the subject had his eyes closed.

The study has also recruited a control population matched for age, sex, handedness and social class of father. Comparison of the patient (Figure 5.8a) with his control reveals a marked reduction of blood flow in the basal ganglia in the unmedicated state. Following the course of neuroleptic medication, not only was there a marked reduction of symptoms (as measured using the BPRS), but also a clear change in the SPECT scan (Figure 5.8b) with increases in the rCBF in the basal ganglia. Such an effect is most likely to be due to the action of the neuroleptic. Group analysis using paired comparisons and also using the z-score method outlined in Chapter 1 is required to analyse the entire data set.

The development of stable radiolabelled ligands which can serve as markers for neurotransmitter subtypes has prompted a further direction of research in schizophrenia. Work to date has focused largely on the dopaminergic D1 and D2 receptor subtypes. This focus of interest is in part due to the long-standing association of dopamine with the pathophysiology of schizophrenia and, more pragmatically, due to the absence of a stable radioligand for the serotonin system receptors. Much SPECT research has drawn heavily on previous studies using PET, but receptor binding studies are now being reported. The development of this field may well have both research and clinical implications, since the level of receptor occupancy may be significant in terms of the likelihood of continued health or relapse. Nordström and colleagues at the Karolinska Institute, Stockholm compared levels of receptor density in schizophrenic patients and healthy controls and failed to find any difference in density between the two groups, but other centres have found increased D2 receptor densities in drug-naive schizophrenic patients. The availability of receptor sites to marking ligands also enables assessment of the level of receptor binding by neuroleptic drugs, an assessment which might be useful in determining drug dosage. Most of this work has been undertaken using PET rather than SPECT, but with the recent development of dopamine receptor ligands suitable for SPECT (see Chapter 1) this preponderance of PET is now less marked. A most interesting finding has been the high levels of receptor occupancy seen using conventional doses of neuroleptics. Farde et al.[55] demonstrated D2 occupancy levels of 70–89% in individuals receiving chlorpromazine 200 mg/day or haloperidol 12 mg/day. Patients with acute extrapyramidal side effects had higher levels of D2 occupancy. The time course of D2 occupancy following neuroleptic prescription has also been monitored by this group: 70–80% receptor occupancy was reported within 3 h of administration of 4 mg oral haloperidol. Akathisia was reported during the period of maximal D2 receptor blockade.

Studies of the atypical neuroleptic clozapine have shown its profile of receptor occupancy to be quite distinct from that of other more classical neuroleptics. At therapeutically effective doses of clozapine, D2 receptor occupancy was found to be 38–63%, while higher levels of D1 and 5-HT2 receptor occupancy were found (haloperidol and chlorpromazine have negligible levels of D1 receptor binding). Measurement of receptor occupancy levels may provide one method of monitoring the reduction of depot neuroleptic prescribing to the minimum level required for treatment while avoiding the risk of relapse. Monitoring of the rate of receptor occupancy following neuroleptic administration has shown rapid high levels of occupancy at modest dose levels, suggesting that symptom resolution depends on a mode of action more complex than simple dopamine receptor blockade. Nevertheless, given that the risk of developing long-term movement disorder is thought to be dependent (to some extent) on the total lifetime exposure to neuroleptic, such measurement could significantly alter clinical practice.

Affective disorders

In comparison with schizophrenia, the functional imaging of the affective disorders has not been the focus of such sustained and intense activity. The relapsing nature of the unipolar form of these conditions, with episodes of depression interspersed by periods of recovery and normal functioning, make them ideal for study using SPECT, since each subject can serve as his own matched control. Several studies have carried out serial SPECT scans at times of relapse and recovery. Fewer studies have been undertaken in patients with the bipolar form of the affective disorders. Such patients may suffer rapid and extreme changes in mood between periods of mania and periods of depression.

Initial SPECT studies concentrated on depression, imaging cohorts of depressed patients matched by normal controls rather than waiting for patients to recover and be rescanned. The design of these studies was based around the methodology and findings of PET studies which had revealed a similar pattern to that seen in schizophrenia, with a loss of the tracer uptake gradient between anterior and posterior cortical regions, reflecting relative hypofrontality of tracer uptake. In a SPECT study, Austin et al.[56] demonstrated reduced cortical and subcortical uptake of radioisotope in a control-matched depressive population. Most significant impairment of uptake was in the temporal, parietal and inferior frontal areas. The severity of depression (measured using the Hamilton rating scale) revealed a negative association with anterior uptake. Interestingly, an association between depressive psychotic symptomatology (measured using the Newcastle scale) and increased tracer uptake in the cingulate area and frontal cortex was found, the latter being similar to the pattern seen in drug-induced and positive-symptom schizophrenic psychotic states. These findings using SPECT techniques confirm previous results using PET[57] in which reduced rCBF was demonstrated in left anterior cingulate and left dorso-lateral prefrontal cortex. Assessment by repeat scanning after recovery has demonstrated increased uptake in basal ganglia and inferior anterior cingulate cortex, suggesting that state changes in depressive illness are confined to inferior limbic and subcortical regions[58].

The chronic effects of relapsing depressive illness on regional blood flow have

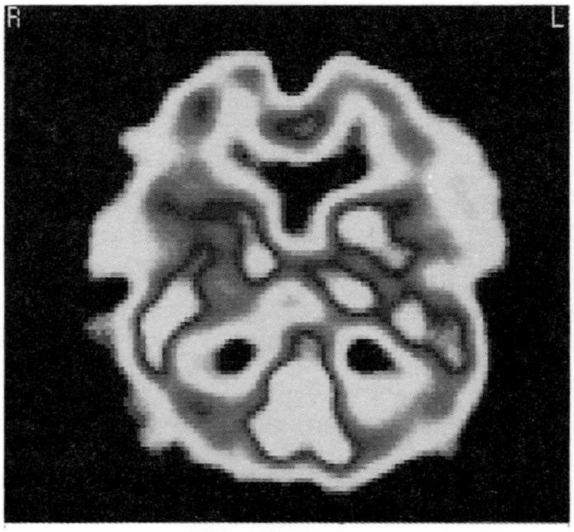

Figure 5.9. This SPECT scan of a rapid cycling manic-depressive male subject aged 45. An axial section at the level of the basal ganglia is presented. The patient has a treatment-resistant, rapidly fluctuating mood disorder which swings from mania to depression and back over a 2 month period. Despite treatment with a variety of drugs, these mood swings have not been successfully stabilised, and have become increasingly severe over the 20 year course of the illness. At the time of the scan, the subject was in a depressed state (scoring 22 on the Beck Depression Inventory) and the scan was performed in a resting state. The features worthy of note are marked reduction in bilateral frontal perfusion, with a prominent reduction of cortex as evident by the widened frontal inter-hemispheric space. Such findings are consistent with those of reports published by other centres, as part of SPECT studies of depressive subjects.

yet to be published: the ideal prospective study would follow patients through successive relapses of illness. It is possible that areas of sustained hypoperfusion remain after clinical recovery, and such findings may be predictive of future relapse. We have found evidence of frontal hypoperfusion and frontal atrophy in subjects with rapid cycling manic and depressive mood swings (Figure 5.9). Subjects with such severe clinical forms of the disorder warrant particular study, since the rapid fluctuation of mood state permits within-subject comparisons of rCBF and objective rating of affective state to be made over a relatively short time.

The treatment methods employed in depression are yet to be fully studied using SPECT techniques. The variety of antidepressants available and their differing profiles of action on neurotransmitter systems, together with inter-patient differences in response, have long intrigued psychiatrists. The recent development of selective serotinergic reuptake inhibitions has focused attention on the 5HT system in depression, but the use of receptor binding ligands to answer some of these questions remains problematic in the absence of a stable 5HT SPECT agent. When such a substance becomes available, this will be a keenly contested research area in psychiatry. Such studies might, for example, lead to clinical tests of receptor density and functioning which can enable a rational choice of antidepressant therapy.

A number of non-drug treatments are employed in the management of depression, varying between electroconvulsive therapy at one extreme and cognitive behavioural therapy at the other. The overall change in rCBF following a course of such treatments can be studied using SPECT and changes specific to a particular treatment can therefore be measured. Studies of this kind are yet to be published. In the not too distant future measurement of receptor density and occupancy following different treatment modalities will also be possible.

One treatment for depression which has rather passed out of vogue is that of sleep deprivation. An interesting treatment study evaluating the effects of sleep deprivation on limbic and frontal lobe activity in depression has been reported by Ebert et al.[59]. All subjects in this study were drug-free, suffered from major depression and showed hypoperfusion in the left anterolateral prefrontal cortex on initial SPECT scan. Following sleep deprivation, responders (50% of the group) reported a reduction in depressive symptomatology. Changes in rCBF were found only in this responder group, with a small increase of relative perfusion in the upper parts of the anterolateral prefrontal cortex. Retrospectively, the responders were found to have pretreatment hyperperfusion of the fronto-orbital cortex, hippocampus, parahippocampus amygdala and inferior lateral temporal lobe. After sleep deprivation the responders showed a return to normal levels of perfusion in these areas.

Neuroses

Obsessive compulsive disorder (OCD) is rapidly recognized as a distinct discrete condition, separate from the neuroses. As a psychiatric condition, OCD is characterized by recurrent, unwanted, distressing thoughts (obsessions) and conscious, ritualized acts (compulsions) which have often arisen as attempts to deal with the increasing levels of anxiety generated by the obsessions. The current treatments of choice for these conditions vary between the use of pharmacotherapy, in particular serotonin reuptake inhibitors, and behaviour therapies employing exposure and response-prevention techniques. Evidence for an underlying abnormality of brain function is suggested by the strong association between OCD and similar behavioural problems in a number of neurological conditions. The presence of neurological 'soft signs' and the use of psychosurgery in the treatment of this condition has further focused attention on the frontal lobes and basal ganglia. Recently SPECT researchers have turned their attention to OCD. While ^{133}Xe SPECT has not shown any difference between OCD and control groups[60], a ^{99}Tc^{m} HMPAO SPECT study showed OCD patients to have increased uptake in the high dorsal parietal cortex and orbital frontal cortex area bilaterally, and increased uptake in left posterofrontal cortex. The OCD patients also had significantly reduced [^{99}Tc^{m}]HMPAO uptake bilaterally in the head of the caudate nucleus. Such findings are consistent with earlier reports using PET techniques[61].

Other neurotic conditions tend to have less specific psychopathologies, with an overall prominence of anxiety in all conditions. SPECT studies of panic attacks (the most severe manifestation of anxiety) have yielded consistent results. Administration of yohimbine (an alpha2-adrenergic antagonist) has been used to induce a panic attack in patients with panic disorder: this agent reduces blood flow in the frontal lobes of such patients[62] while failing to produce a similar effect in normal controls. No studies have investigated specific neurotic conditions. There is perhaps less of an expectation for specific areas of abnormality dependent upon diagnosis, as rCBF patterns are likely to reflect the non-specific effects of anxiety experienced by such patients.

Substance abuse

This remains a complex area, since drug abusers often use a variety of illicit substances depending on availability, and some also have psychiatric illness. Nevertheless, the effects of particular drugs have been of interest to researchers, since SPECT provides a method of objective measurement of the site of action and associated subjective experiences reported by those exposed to the drug. Cocaine abusers have been reported to show decreased relative CBF in the prefrontal cortex when studied with PET. Such abnormalities remained after cocaine withdrawal[63]. Non-specific effects of intoxication, whether by cocaine, morphine or alcohol, reveal a reduction in global glucose utilization when investigated by PET.

SPECT research has largely focused on volunteer groups taking psychoactive substances in controlled conditions. Mescaline induces a brief psychotic state with specific effects on the visual system. SPECT imaging of this state revealed a hyperfrontal pattern most marked in the right hemisphere. A similar study using amphetamine in a population of chronic stable schizophrenics maintained by haloperidol has been reported by Daniel et al.[64]. The study used a sensorimotor task control state and the WCS activation test. The control state showed a slight global reduction in rCBF. The WCS task demonstrated a marked change between placebo and amphetamine states, with activation of the left DLPFC and an associated improvement in task performance.

SPECT clearly offers the opportunity to study in much greater depth the effects of psychoactive drugs on brain function, as well as to gain greater understanding of the disease processes which have so long remained a mystery in psychiatry.

Conclusions

This chapter has examined the use of SPECT in single disease entities and its ability to discriminate between different disease processes, emphasizing the sensitivity of the technique to changes in arousal, age, the presence of psychoactive substances and the type of task performed by the subject at the time of investigation. In some areas, these abilities already allow the use of SPECT as an adjunct to the investigatory options available in assessing patients. In dementia, SPECT imaging of cerebral blood flow distribution is of established diagnostic value in separating Alzheimer's disease from vascular dementia, other dementias and depression, and enables measurement to be made of the alterations in regional perfusion pattern produced by cognitive enhancing drugs. Guidelines issued to doctors by the Royal College of Radiologists in the UK listed SPECT as the first line imaging investigation in dementia, assisted by MRI to improve specificity in the differential diagnosis. The scope for the clinical use of SPECT in psychiatry may increase, allowing for more accurate predictions for recovery and planning of suitable rehabilitation to be established at an earlier stage in treatment.

At present the vast majority of SPECT investigations use a cerebral blood flow radiopharmaceutical. Experience with specific neuroreceptor radioligands is growing, using commercially available agents such as IBZM to study the D2 system in schizophrenia or iomazenil to study the GABA system in dementia, or using 'in-house' products such as [^{123}I]QNB to measure muscarinic cholinergic receptor density in Alzheimer's disease.

SPECT is showing potential for investigating other systems which may be involved in psychiatric disorders, in particular the GABA and glutamate systems, while studies in schizophrenia using SPECT with ^{133}Xe have confirmed the association of hypofrontality with negative symptoms and several studies using a variety of techniques have demonstrated abnormal functional activation on cognitive task performance. Studies on affective disorders, neuroses and substance abuse are adding to current knowledge of regional functional disturbance in those conditions and of the extent to which this is reversed in response to treatment.

References

1. Roth M, Wischik CM, Evans N, Mountjoy CQ. Convergence and cohesion of recent neurobiological findings in relation to Alzheimer's disease and their bearing on its aetiological basis. In: Bergener M, Ermine M, Stahelim HB, editors. Thresholds in ageing. London: Academic Press, 1985:117–46.
2. Robins LN, Helzer JE, Weissman MN. Lifetime prevalence of specific psychiatric disorders in three sites. Arch Gen Psychiatry. 1984;41:949–58.
3. Weissman MN, Leaf PJ, Tischler GL. Affective disorders in five United States communities. Psychol Med. 1988;18:141–53.
4. Regier DA, Boyd JH, Burke JD. One month prevalence of mental disorders in the United States. Arch Gen Psychiatry. 1988;45:977–86.
5. Gur RC, Erwin RJ, Gur RE. Neurobehavioural probes for physiologic neuroimaging studies. Arch Gen Psychiatry. 1992;49:409–14.
6. Ginsberg MD, Chang JY, Kelley RE, et al. Increases in both cerebral glucose utilization and blood flow during execution of a somatosensory task. Ann Neurol. 1988;23:152–60.
7. Reivich M, Gur R, Alavi A. Positron emission tomographic studies of sensory stimuli, cognitive processes and anxiety. Human Neurobiol. 1983;2:25–33.
8. Gur RC, Gur RE, Resnick SM, Skolnick BE. The effect of anxiety on cortical cerebral blood flow and metabolism. J Cereb Blood Flow Metab. 1987;7:173–7.
9. Gur RC, Gur RE, Obrist WD, Skolnick BE, Reivich M. Age and regional cerebral blood flow at rest and during cognitive activity. ARch Gen Psychiatry. 1987;44:617–21.
10. Bartlett EJ, Wolkin A, Brodie JD, Laska EM, Wolf AP, Sanfilipo M. Importance of pharmacologic control in PET studies: effects of thiothixene and haloperidol on cerebral glucose utilisation in chronic schizophrenia. Psychiatry Res. 1991;40:115–24.
11. Hermle L, Funfgeld M, Oepen G, et al. Mescaline-induced psychopathological, neuro-

psychological, and neurometabolic effects in normal subjects: experimental psychosis as a tool for psychiatric research. Biol Psychiatry. 1992;32:976–91.
12. Ingvar DH, Franzen G. Distribution of cerebral activity in chronic schizophrenia. Lancet. 1974;1484–6.
13. Wyper D, Teasdale E, Patterson J, et al. Abnormalities in rCBF and computed tomography in patients with Alzheimer's disease and in controls. Br J Radiol. 1993;66:23–7.
14. Coyle JT, Price DL, De Long MR. Alzheimer's disease: a disorder of cortical cholinergic innervation. Science. 1983;219:1184–90.
15. Blessed G. Clinical aspects of the senile dementias. In: Roberts PJ, editor. Biochemistry of Dementia. Chichester: John Wiley and Sons, 1980:1–14.
16. Sim M, Turner E, Smith WT. Cerebral biopsy in the investigation of presenile dementia: I. clinical aspects. Br J Psychiatry. 1966;112:119–25.
17. Goldman-Rakic PS. Modular organisation of prefrontal cortex. Trends Neurosci. 1984;7:419–29.
18. Roth M, Tym E, Mountjoy CW, et al. CAMDEX: a standardised instrument for the diagnosis of mental disorder in the elderly with special reference to the early detection of dementia. Br J Psychiatry. 1986;149:698–709.
19. Sourander P, Sjogren H. The concept of Alzheimer's disease and its clinical implications. In: Wolstenholme GEW, O'Connor M, editors. Alzheimer's disease. London: Churchill, 1970: 11–36.
20. Rossor MN, Iversen LL, Reynolds GP, Mountjoy CQ, Roth M. Neurochemical characteristics of early and late onset Alzheimer's disease. Br Med J. 1984;288:961–4.
21. Bonte FJ, Ross ED, Chehabi HH, Devous MD. SPECT study of regional cerebral blood flow in Alzheimer's disease. J Comput Assist Tomogr. 1986;10:579–83.
22. Cohen MB, Graham LS, Lake R, et al. Diagnosis of Alzheimer's disease and multi-infarct dementia by tomographic imaging of iodine-123IMP. J Nucl Med. 1986;27:769–74.
23. Sharp P, Gemmell H, Cherryman G, Besson J, Crawford J, Smith F. Application of iodine-123-labelled isopropylamphetamine imaging to the study of dementia. J Nucl Med. 1986;27:761–8.
24. Johnson KA, Mueller ST, Walshe TM, English RJ, Holman BL. Cerebral perfusion imaging in Alzheimer's disease. Arch Neurol. 1987;44:165–8.
25. Jagust WJ, Budinger TF, Reed BR. Diagnosis of dementia with single photon emission computed tomography. Arch Neurol. 1987;44:258–62.
26. Andersen AR, Friberg HH, Schmidt JF, Hasselbalch SG. quantitative measurements of cerebral blood flow using SPECT and [^{99m}Tc]-d,1-HMPAO compared to xenon-133. J Cerebr Blood Flow Metab. 1988;8:S69–S81.
27. HunterR, McLuskie R, Wyper D, et al. The pattern of function-related regional cerebral blood flow investigated by single photon emission tomography with 99mTc-HMPAO in patients with presenile Alzheimer's disease and Korsakoff's psychosis. Psychol Med. 1989;19:847–55.
28. Neary D, Snowden JS, Shields RA, et al. Single photon emission tomography using ^{99m}Tc-HMPAO in the investigation of dementia. J Neurol Neurosurg Psychiatry. 1987;50:1101–9.
29. Gemmell HG, Sharp PF, Besson JAO, et al. Differential diagnosis in dementia using the cerebral blood flow agent ^{99m}Tc HMPAO: a SPECT study. J Comput Assist Tomogr. 1987;11: 398–402.
30. Burns A, Philpot MP, Costa DC, Ell PJ, Levy R. The investigation of Alzheimer's disease with single photon emission tomography. J Neurol Neurosurg Psychiatry. 1989;52:248–53.
31. Holman BL, Johnson KA, Gerada B, Carvalho PA, Satlin A. The scintigraphic appearance of Alzheimer's disease: a prospective study using technetium-99m-HMPAO SPECT [published erratum appears in J Nucl Med 1992 Apr; 33(4):484] [see comments]. J Nucl Med. 1992;33: 181–5.
32. Jobst KA, Hindley NJ, King E, Smith AD. The diagnosis of Alzheimer's disease: a question of image? J Clin Psychiatry. 1994;55:22–31.
33. Freedman LS. Evaluating and comparing imaging techniques: a review and classification of study designs. Br J Radiol. 1987;60:1071–81.
34. Haxby JV, Grady CL, Koss E, et al. Heterogeneous anterior-posterior metabolic patterns in dementia of the Alzheimer type. Neurology. 1988;38:1853–63.

35. Montaldi D, Brooks DN, McColl JH, et al. Measurements of regional cerebral blood flow and cognitive performance in Alzheimer's disease. J Neurol Neurosurg Psychiatry. 1990;53:33–8.
36. Goldenberg G, Podreka I, Suess E, Deecke L. The cerebral localization of neuropsychological impairment in Alzheimer's disease: a SPECT study. J Neurol. 1989;236:131–8.
37. Owens J, Murray T, McCulloch J, Wyper D. Synthesis of (R,R) 123I-QNB, a SPECT imaging agent for cerebral muscarinic acetylcholine receptors in vivo. J Labelled Compounds Radiopharmaceuticals. 1991;31:45–60.
38. Weinberger DR, Gibson R, Coppola R, et al. The distribution of cerebral muscarinic acetylcholine receptors in vivo in patients with dementia. A controlled study with 123IQNB and single photon emission computed tomography. Arch Neurol. 1991;48:169–76.
39. Weinberger DR, Jones D, Reba RC, et al. A comparison of FDG PET and IQNB SPECT in normal subjects and in patients with dementia. J Neuropsychiatry Clin Neurosci. 1992;4: 239–48.
40. Wyper DJ, Brown D, Patterson J, et al. Deficits in iodine-labelled 3-quinuclidinyl binding in relation to cerebral blood flow in patients with Alzheimer's disease. Eur J Nucl Med. 1993; 20:5:379–86.
41. Probst A, Cortes R, Ulrich J, Palacios JM. Differential modification of muscarinic cholinergic receptors in the hippocampus of patients with Alzheimer's disease: an autoradiographic study. Brain Res. 1988;450:190–201.
42. Geaney DP, Soper N, Shepstone BJ, Cowen PJ. Effects of central cholinergic stimulation on regional cerebral blood flow in Alzheimer disease. Lancet. 1990;335(i):1484–7.
43. Hunter R, Wyper DJ, Patterson J, Hansen MT, Goodwin GM. Cerebral pharmacodynamics of physostigmine in Alzheimer's disease investigated using single-photon computerised tomography. Br J Psychiatry. 1991;158:351–7.
44. Ebmeier KP, Hunter R, Curran SM, et al. Effects of a single dose of the acetylcholinesterase inhibitor velnacrine on recognition memory and regional cerebral blood flow in Alzheimer's disease. Psychopharmacology (Berl). 1991;108:103–9.
45. Hunter R, Brown D, Bach L, Wyper D, Patterson J, McCulloch J. The effect of the acetylcholinesterase inhibitor velnacrine on cognition and regional cerebral blood flow in Alzheimer's disease. In: Corain B, Iqbal K, Nicolini M, Winbald B, Wisniewski H, Zatta P, editors. Alzheimer's Disease: Advances in clinical and basic research. Chichester: John Wiley, 1993:569–75.
46. American Psychiatric Association. Diagnostic and statistical manual of mental disorders, 3rd rev. edn. Washington, DC: American Psychiatric Association, 1987.
47. McKhann G, Drachman D, Folstein M. Clinical diagnosis of Alzheimer's disease: report of NINCDS-ADRDA work group under the auspices of Department of Health and Human Service task force on Alzheimer's disease. Neurology. 1984;34:939–44.
48. Hardy J, Cowburn R, Barton A, Reynolds G. A disorder of cortical GABAergic innervation in Alzheimer's disease. Neurosci Lett. 1987;73:192–6.
49. Schubiger PA, Hasler PH, Beer-Wohlfahrt H. Evaluation of a multicentre study with iomazenil – a benzodiazepine receptor ligand. Nucl Med Commun. 1991;12:569–82.
50. Sheppard G, Gruzelier J, Manchanda R, et al. Positron emission tomographic scanning in predominantly never-treated acute schizophrenic patients. Lancet. 1983:1448–52.
51. Andreasen NC, Rezai K, Alliger R, et al. Hypofrontality in neuroleptic-naive patients and in patients with chronic schizophrenia. Assessment with xenon 133 single-photon emission computed tomography and the Tower of London. Arch Gen Psychiatry. 1992;49:943–58.
52. Paulman RG, Devous MDS, Gregory RR, et al. Hypofrontality and cognitive impairment in schizophrenia: Dynamic single-photon tomography and neuropsychological assessment of schizophrenic brain function. Biol Psychiatry. 1990;27:377–99.
53. Weinberger DR, Berman KF, Ilowsky BP. Physiological dysfunction of dorsolateral prefrontal cortex in schizophrenia. Arch Gen Psychiatry. 1988;45:609–15.
54. Lewis SW, Ford RA, Syed GM, Reveley AM, Toone BK. A controlled study of ^{99m}Tc-HMPAO single photon emission imaging in chronic schizophrenia. Psychol Med. 1992;22:27–35.
55. Farde L, Nordstrom A-L, Wiesel F-A, Pauli S, Halldin C, Sedvall G. Positron emission

tomographic analysis of central D1 and D2 dopamine receptor occupancy in patients treated with classical neuroleptics and clozapine. Arch Gen Psychiatry. 1992;49:538–44.

56. Austin MP, Dougall N, Ross M, et al. Single photon emission tomography with 99mTc-exametazime in major depression and the pattern of brain activity underlying the psychotic/neurotic continuum. J Affect Disord. 1992;26:31–43.
57. Bench CJ, Friston KJ, Brown RG, Scott LC, Frackowiak RSJ, Dolan RJ. The anatomy of melancholia – focal abnormalities of cerebral blood flow in major depression. Psychol Med. 1992;22:607–15.
58. Goodwin GM, Austin MP, Dougall N, et al. State changes in brain activity shown by the uptake of ^{99m}Tc-exametazime with single photon emission tomography in major depression before and after treatment. J Affect Disord. 1993;29:243–53.
59. Ebert D, Feistel H, Barocka A. Effects of sleep deprivation on the limbic system and the frontal lobes in affective disorders: a study with Tc-99m-HMPAO SPECT. Psychiatry Res. 1991;40: 247–51.
60. Rubin RT, Villanueva Meyer J, Ananth J, Trajmar PG, Mena I. Regional xenon 133 cerebral blood flow and cerebral technetium 99m HMPAO uptake in unmedicated patients with obsessive-compulsive disorder and matched normal control subjects. Determination by high-resolution and single-photon emission computed tomography. Arch Gen Psychiatry. 1992;49:695–702.
61. Baxeter LR, Schwartz JM, Bergman KS, et al. Caudate glucose metabolic rate changes with both drug and behavior therapy for obsessive-compulsive disorder. Arch Gen Psychiatry. 1992;49:681–9.
62. Woods SW, Koster K, Krystal JK. Yohimbine alters regional cerebral blood flow in panic disorder. Lancet. 1988;11:678–0.
63. Volkow ND, Mullani N, Gould KL, Adler S, Krajewski K. Cerebral blood flow in chronic cocaine users: a study with positron emission tomography. Br J Psychiatry. 1988;152:641–8.
64. Daniel DG, Weinberger DR, Jones DW, et al. The effect of amphetamine on regional cerebral blood flow during cognitive activation in schizophrenia. J Neurosci. 1991;11:1907–17.

6. The use of SPECT in the analysis of brain tumours

GARTH S. CRUICKSHANK

Introduction

Despite advances in the treatment and prognosis of many solid tumours over the last 20 years, the outcome for adults with malignant primary brain tumours remains very poor. The incidence is reported at approximately 6–7 per 100 000 of the population. The West of Scotland (the catchment areas of the Glasgow Institute of Neurological Sciences) has a population of 2.7 million. This means that 120 to 150 new cases present each year.

Glioblastoma multiforme is a lethal malignancy. Subtotal surgical resection and post-operative radiotherapy palliate symptoms and prolong survival, but very few patients survive more than 3 years following diagnosis. More than half of these patients will die within one year, despite optimal therapy. The probability of survival with anaplastic astrocytoma is slightly better, with 15–20% of patients alive 5 years following combined modality treatment, as shown in several studies[1]. Of those patients presenting with intermediate and low grade astrocytic lesions over 50% will develop a high grade malignancy, with its associated morbidity and mortality, in a 5-year period[2]. Patients with cerebral metastases from a number of different sources have a varying survival with between 30–50% of patients surviving one year. A few patients, such as those with breast disease and chorionic carcinoma, may live for many years.

In randomized studies, postoperative radiotherapy is the only modality of treatment to give significantly prolonged median survival[3]. Chemotherapy trials have suggested some small survival advantage in the highly malignant tumours[4]. Evidence over the years has suggested that reduction in tumour size by surgery results not only in improvement in symptoms but also in prolonged survival as well as an improved result from radiotherapy[5]. The pattern of treatment failure is predominantly local recurrence, over 80% of patients having recurrent disease in the high-dose target region of the radiation portals[6]. This suggests that primary malignant brain tumours have a unique place in the spectrum of clinical radioresistance: epithelial tumours of the same size are frequently cured locally by comparable dosages of radiation[7].

In cell culture the cytotoxic effect of radiation on squamous cell carcinomas (SCC) originating from the head and neck is similar to that on glial tumours[8], and the dose required to kill 50% of SCC tumours explanted into mice is similar to that required for the same killing effect in explanted glial tumours[9]. However, we

R. Duncan (ed): SPECT Imaging of the Brain, 161–178.

know that in vivo SCC tumours require a lower dose for control than do glial tumours[10]. This suggests that radioresistance is not completely intrinsic to the tumour cell, and may be influenced by factors in the cellular environment. This may be important with respect to the development of tests directed towards differentiating resistant and non-resistant areas of the tumours.

Despite nearly three decades of intense research, conventional modalities of surgery, chemotherapy and external beam radiation therapy have had relatively little impact on the survival of patients with malignant brain tumours. There is little doubt that local surgical treatment is inadequate, as recurrences arise from the edge of the operated area. Such residual tumour requires more generalized treatment, which cannot necessarily be achieved by radiation. While we await developments in such fields as enhanced chemotherapy and gene transfer, it is clear from a number of studies that imaging techniques directed at detecting and defining areas of viable tumour, clonogenic and non-clonogenic, can markedly improve our strategies for applying current and developing treatment modalities.

SPECT used with appropriate radioligands offers, by virtue of selective uptake only by viable tumour cells, the possibility of imaging the viable portion of a given tumour[10]. This is an important concept to consider. Most tumour assessment has been based on evaluation of CT and MRI scans, which primarily give morphological information. Brain tumours are notably heterogeneous and consist of live growing cells, dormant viable cells, a small variable proportion of inflammatory cells, dead cells and areas of necrosis and profound oedema[11]. All of these factors contribute to the image of the 'lesion' seen on conventional scanning. Their appearances can be further complicated by the distorting influences of mass effect and ischaemia. The use of contrast agents helps to delineate areas of blood–brain barrier disruption, although this may be determined by vascular abnormality rather than by tumour mass. The formation of cysts by astrocytomas and some metastatic tumours confuses this kind of assessment of tumour volume, making a true assessment of surgical impact or treatment response very difficult. Inflammation, blood–brain barrier breakdown and swelling are also associated with radiation necrosis and may also further exacerbate the problem. SPECT offers the ability to show only that part of the tumour that is actually live and growing, and consequently the change to much more directed forms of therapy becomes possible.

Studies using BrdU (radiolabelled uridine) to look at the portion of tumours that are actually proliferating have shown a labelling index of only 0.3–10.0% in fresh tumour samples[12]. This implies that compounds taken up only by proliferating cells will only indicate that fraction of the total volume of the tumour that is viable, i.e. not showing the considerable fraction of the tumour that is alive but unproliferating This is important, as current treatments are directed at the proliferating fraction only. The ultimate aim of treatment should be nothing less than the total killing of the viable tumour mass, particularly since dormant cells may already be transformed and become proliferative at a later stage.

SPECT imaging may offer other important benefits. It can be argued that waiting for symptomatic neurological deterioration risks injury to brain substance. Could SPECT detect recurrence before further irreparable brain damage occurs? A technique with the ability to selectively image brain tumour tissue could detect regrowth early, and exclude many of the compounding factors (e.g. oedema, necrosis) that make treatment evaluation so difficult, so that the tumour response itself would become the central issue. Such a technique must be reliable, with usable demarcation levels. It must be acceptable to patients and doctors alike. It should not be expensive and should not interfere with treatment, except to advantage. Some SPECT techniques come close to answering many of these needs.

Imaging of brain tumours using ^{201}Tl

A number of different agents have been used in an attempt to selectively image brain tumours. This chapter is intended to provide a guide to the evaluation of agents in common use, and emphasizes clinically relevant factors to consider when using them.

Any isotope used for three-dimensional imaging of tumours should have a very high affinity for neoplastic tissue relative to the surrounding normal tissue. Most of the intrinsic brain tumours develop within a matrix of white matter. It is generally accepted that the more malignant a brain tumour becomes, the more damaged the blood–brain barrier can be. This is important when evaluating compounds that must traverse the blood–brain barrier to gain access to the tumour.

^{201}Tl was first recognized as a marker for tumours when studies were performed looking at its use in assessing cardiac muscle viability[13]. A number of workers noticed that lung tumours could be identified by thallium coincidentally while imaging cardiac activity[14]. Subsequently, a more systematic evaluation of uptake by tumours showed that a number of intrinsic and metastatic lesions could be identified within the cranial cavity[15]. A number of studies directed to the use of thallium in the diagnosis of cerebral lesions revealed a substantial uptake by primary and some metastatic cerebral tumours. The uptake by normal brain was shown to be minimal[16] (Figure 6.1). Three major factors have been identified by a number of authors as being responsible for determining isotopic concentration by tumours in the brain.

Blood flow

A number of studies looking at distribution of blood flow using HMPAO have shown enormous variation in flow to different tumour regions. An adequate blood flow to the tumour is presumably necessary for isotope delivery but, in a number of studies, complete absence of HMPAO uptake has not precluded concentration of thallium in malignant areas (Figure 6.2; see also colour section). It therefore seems fair to assume that there is a threshold blood flow level above which thallium can gain access to tumours and that this threshold is by and large achieved in the intracranial cavity[15].

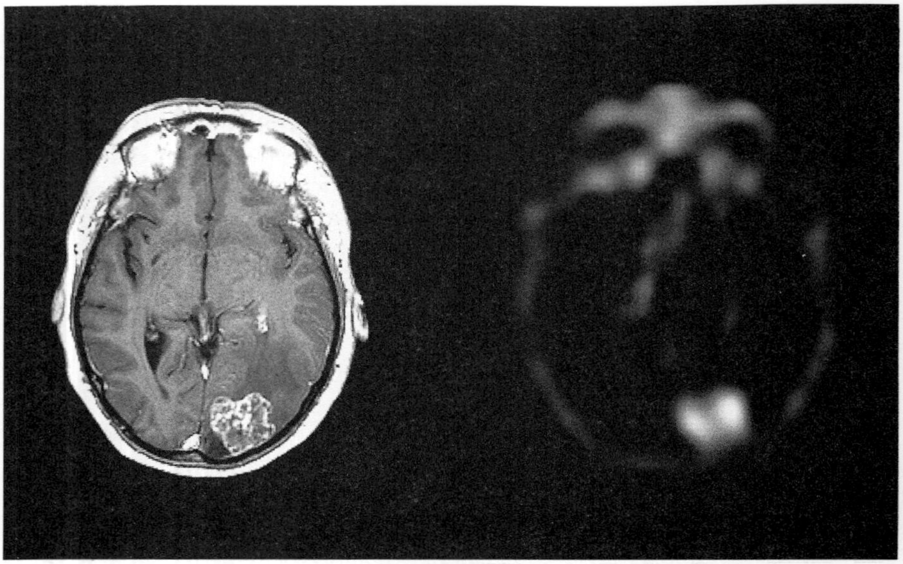

Figure 6.1. (Left) Gadolinium MRI image, showing an enhancing tumour in the left occipital lobe. (Right) Corresponding slice of the coregistered thallium image, showing thallium uptake corresponding to the enhancing area of the tumour.

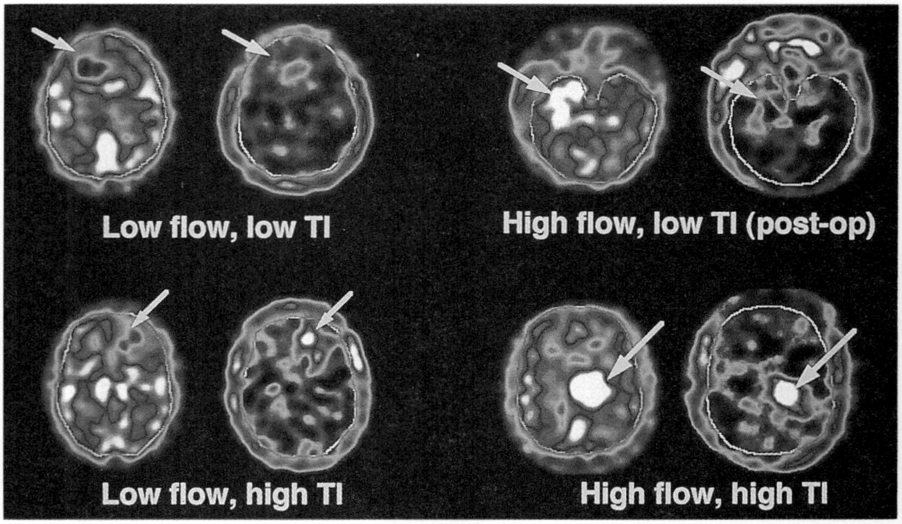

Figure 6.2. Four combinations of blood flow (demonstrated by HMPAO) and thallium uptake by tumours. Of each pair of images, the HMPAO image is on the left, thallium on the right. There does not appear any clear cut relationship between the two. (For colour version of this figure see page 185.)

The uptake of ^{201}Tl in the brain reaches a maximum at 5–10 min after i.v. injection. This relatively slow uptake is thought to be related to heterogeneity of cell type within the tumour[16]. Washout curves are quite variable and there is some evidence that retention of isotope is related to proliferative potential[18]. While variations in tissue perfusion do not necessarily correlate with thallium findings, it has been suggested that they can be useful in differentiating equivocal findings with thallium, such as in the problem of identifying radiation necrosis[19].

Blood–brain barrier disruption

Increased blood–brain barrier permeability alone will not result in increased thallium uptake, since non-neoplastic areas of increased blood–brain barrier breakdown, e.g. that associated with areas of radiation necrosis, may show little or no ^{201}Tl uptake[15]. Non-specific leakage in oedematous tissue has not been demonstrated with thallium: evidence from several authors confirms that uptake into areas of abscess and cerebritis is low and does not interfere with tumour assessments. Even in the immediate post-operative period, when local tissue disruption is inevitable, thallium uptake only seems to occur where high-grade malignant tumour is present. A systematic evaluation of the relationship between blood–brain barrier permeability and thallium uptake (e.g. using autoradiography) has not yet been carried out[20]. Specific uptake of thallium tends in any event to be concentrated in malignant brain tumours, which are themselves associated with blood–brain barrier disruption; it is therefore not yet possible to say that thallium uptake is not in some way related to that disruption. It might be that both blood–brain barrier disruption and specific tumour cell uptake are required, and there is certainly support in the literature for an uptake mechanism[17,21]; see below.

Cell density/active uptake

The third factor which determines thallium visualization on isotopic scanning can be regarded as a combination of specific uptake activity and cell density. Malignant brain tumour cells are thought to take up thallium by an adenosine triphosphate-dependent cell membrane pump (Na^+,K^+-ATPase)[20]. It has been suggested that membrane movements of thallium and potassium are closely related and that the pump involved cannot differentiate between thallium or potassium. Thallium can compete with potassium inactivation of the Na^+,K^+-ATPase and appears to have an affinity 10 times greater than potassium for a potassium-activating site[22]. Following uptake, retention of thallium by the cell is greatly potentiated over that of potassium. There is significant evidence that Na^+,K^+-ATPase activity is generally increased within malignant brain tumours: specific activity as well as the membrane density of the enzyme is increased. Most tumour cells that have been shown to be hypermetabolic have increased ATPase activity associated with the membrane[23]. One would therefore expect increased thallium uptake to be associated with increased ATPase activity in most malignant tumours.

By definition all malignant brain tumours are associated with increasing cell density (hypercellularity)[24], which itself results in a high concentration of thallium in aggregates of tumour cells. This relates to the technical issue of what activity per unit volume can be resolved using SPECT imaging (see Chapter 1).

A number of other factors might also influence thallium uptake. In cardiac studies there has been shown to be some interaction with the ATPase antagonist digoxin[13]. In brain tumours no such studies have been performed. The most commonly used treatment for brain tumours is steroid (dexamethasone), and there has been at least one report suggesting that steroid use does not influence brain tumour imaging using ^{201}Tl[25]. It is nonetheless important to remember that dexamethasone can affect estimates of tumour volume as a result of its own pharmacological effect, and the use of thallium as a simple marker of viable tumour volume may be influenced by these direct effects. A more rigorous evaluation of this problem is awaited.

In summary, thallium uptake into malignant tumours is not simply a function of blood flow. Blood–brain barrier disruption facilitates thallium entry into brain tumours, and specific uptake is dependent on increased Na^+,K^+-ATPase activity in the membrane of malignant cells, potentiated on the macroscopic scale by increased malignant cell density.

Thallium and the evaluation of malignancy

Grading: The possibility that thallium can be used to differentiate high and low grade lesions has been suggested by a number of authors[17,26]. MRI can provide up to 85% accuracy in the diagnosis of malignant lesions, but considerable evidence is now accumulating that thallium uptake is not only an accurate indicator of malignancy, but can also give a better assessment of malignant potential in equivocal cases than either conventional imaging or histology[27]. Some discrepancies between thallium-based estimates of malignancy and histology can be ascribed to tumour sampling problems.

It should be noted that thallium uptake is not exclusive to malignant glial tumours. Other tumours, including lymphomas, osteosarcomas, and meningiomas, also show varying degrees of uptake. Thallium, therefore, cannot be used to identify cell origin[17]: a positive biopsy-based tissue diagnosis is required. Meningiomas can have a tumour–brain ratio of thallium uptake as high as 5:1, as can some sarcomas and metastases[28]. This can confuse a simple correlation with malignant potential unless the tumour nature is known[16].

Where authors have measured tumour–brain uptake indices, malignancy of intrinsic brain tumours has been shown to correlate with uptake ratios of 2:1. Most evidence points to an index around this figure as diagnostic for malignant behaviour in intrinsic brain tumours. A thallium index of > 1.5 is associated with probable malignancy[22], but whether ratios can predict tumour progression is an interesting question. Preliminary data suggest that, for example, a grade II (WHO) glioma with a thallium index > 1.5 progresses more rapidly than a grade II glioma with a thallium index < 1.5[25]. Further support for the predictive possibilities of thallium SPECT and malignant potential have been reported by Vertosick et al.[29].

The detection of malignant degeneration in low grade tumours: Whenever low grade tumour is confirmed by biopsy and can be followed up by imaging, a change in thallium SPECT scanning from negative to positive can help indicate that malignant change has occurred[22]. This offers some help to the physician in following patients with low grade tumours. It may also indicate at a first presentation which patients might be considered for early surgery. Patients with low grade tumours who undergo radiotherapy are difficult to monitor, since assessment of tumour volume can be unreliable, and tumour response to therapy is largely unmeasurable[30]. This has led to confusion in the literature as to whether aggressive treatment of low grade gliomas is a worthwhile activity. Neuroimaging agents are avidly sought to help with this problem.

Measurement of uptake by indices: Cerebral SPECT is designed to give anatomical information concerning the distribution in brain of the isotope assessed. In practice the information given by the picture seen should not be undervalued in favour of measurement indexes and uptake ratios: the two are complementary. Indices do provide a means of comparing images produced by different laboratories, and workers in the field should evaluate SPECT images produced by different centres before accepting their results. Such a discerning approach is particularly necessary for the surgeon who may use the data for image-directed procedures in the operating room. Common oddities of thallium uptake, such as variable scalp uptake and marked uptake by retro-orbital fat, can be very misleading. Highlighting of the choroid plexus, for example, by MIBI (see below) can handicap assessment of adjacent lesions and make it difficult to differentiate tumour from normal tissue[31].

The selection of regions of interest (ROI) must not be inappropriately prejudiced by morphological data from CT scans. Fitting of a CT–ROI mask to SPECT can lead to erroneously high values for thallium uptake, especially in 'noisy' scans. We do not use CT-determined ROIs in determining ratios, but do use CT to select the largest possible area in the contralateral hemisphere to measure the baseline thallium activity. The most important issue is not so much that an ROI is chosen using CT scanning, versus a signal threshold setting, but that a method is used which enables some degree of internal consistency to be obtained between patients and between measurements. Tumour/brain ratios are more realistic than tumour/heart ratios, and are much more reliable than tumour/scalp ratios, since scalp values are highly variable, particularly in the post-operative period or during radiation therapy.

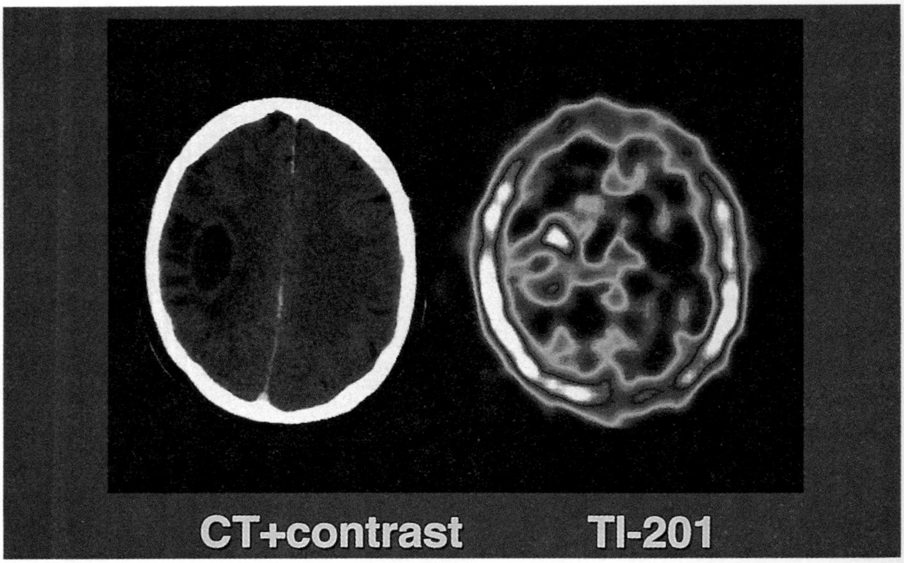

Figure 6.3. CT scan showing a high grade cystic lesion in the right parietal lobe. The simple approach for biopsy would suggest biopsy of the outer wall of the cyst. The SPECT scan shows, however, that the area of maximum thallium activity is in the anterior medial wall of the cyst, and that this is in fact the most sensible target for biopsy. (For colour version of this figure see page 186.)

Practical uses of 201Tl SPECT

Improving biopsy yield: The heterogeneous nature of brain tumours limits the surgeon's ability to direct biopsy at representative viable tissue using conventional imaging as a guide. For example, necrotic tissue may not be distinguished from viable tumour, and its aspiration may indicate malignancy, but not necessarily tissue type. Figure 6.3 (see also colour section) shows a cystic lesion in the right parietal lobe. It can be seen from the accompanying SPECT pictures that conventional biopsy of the most superficial lesion of the tumour would not result in a sample from the most active or most viable portion of the tumour.

The concept of selection of the most viable tissue for biopsy, has been verified using a PET technique[32]. The improvement in yield, together with the reduced risk to the patient, renders the technique particularly useful. If it could be adapted to the much less expensive SPECT technique then the same positive result might be expected[32,33] given suitable image quality. Co-registration of SPECT or PET images with anatomical images is becoming more commonly available[34]. There is little doubt that integration of such a method with the surgical technique would improve biopsy success rate and possibly reduce patient morbidity and mortality. The added benefit of sampling the most diagnostic portion of a tumour is that it is likely to improve our diagnostic precision and assessment of prognostic variables, an assessment which has been hitherto based on variable tissue sampling.

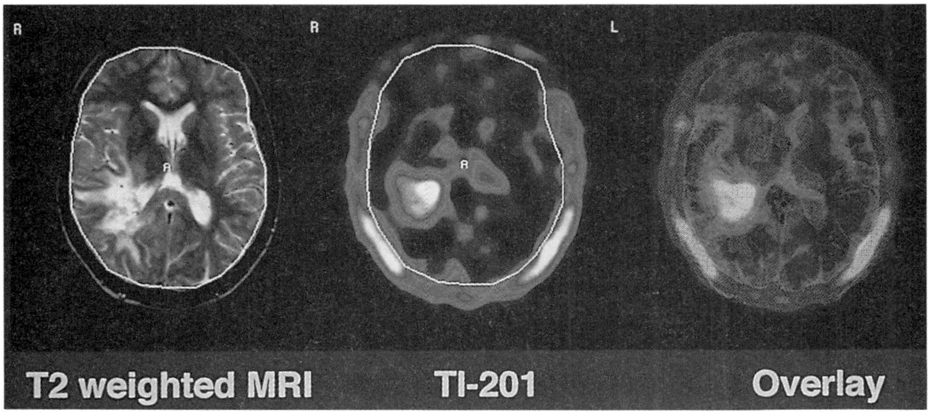

Figure 6.4. MRI scan (left) shows a contrast-enhanced tumour in the left trigonal area. SPECT image (right) co-registered with the MRI so that the limits of the proliferating tumour volume can be determined in terms of MRI stereotaxic coordinates. (For colour version of this figure see page 186.)

Defining the volume for image directed surgery: Recently, Hughes et al. have described a method of co-registering thionine PET images with CT or MRI to aid image-directed brachytherapy[35]. We have used a similar, simpler approach using SPECT and MRI co-registration, to define coordinate limits for tumour resection, using an MRI directed stereotactic resection (Figure 6.4; see also colour section).

The combined data from both MRI and SPECT provides enormously useful information when dealing with the practical issues of operation, from approach trajectory to proximity to eloquent structures. With improved software and technology it should be possible to speed the process of co-registration to the point at which it can be used with for example an ISG wand or a neuronavigator system[36]. Views differ as to whether the quantity of tumour excision can influence survival[37]. As discussed, the volume focal tumour, as determined by MRI and CT, may not be as reliable as that determined by SPECT in deciding what must be removed at surgery. The author's data show that the use of thallium SPECT results in a larger excision, and this is currently being evaluated for survival benefit. The location of the tumour removed can be assessed objectively, and early evidence suggests that absence of post-operative uptake of thallium is associated with prolonged disease-free survival. Whether such a result will be significant in the long term remains to be seen.

Determining the efficacy of surgery: The lack of influence of direct effect of operative disturbance on thallium uptake renders thallium SPECT enormously useful for evaluating the completeness of operative tumour removal (Figure 6.5; see also colour section).

The ability to measure cytoreductive effect in patients will help considerably

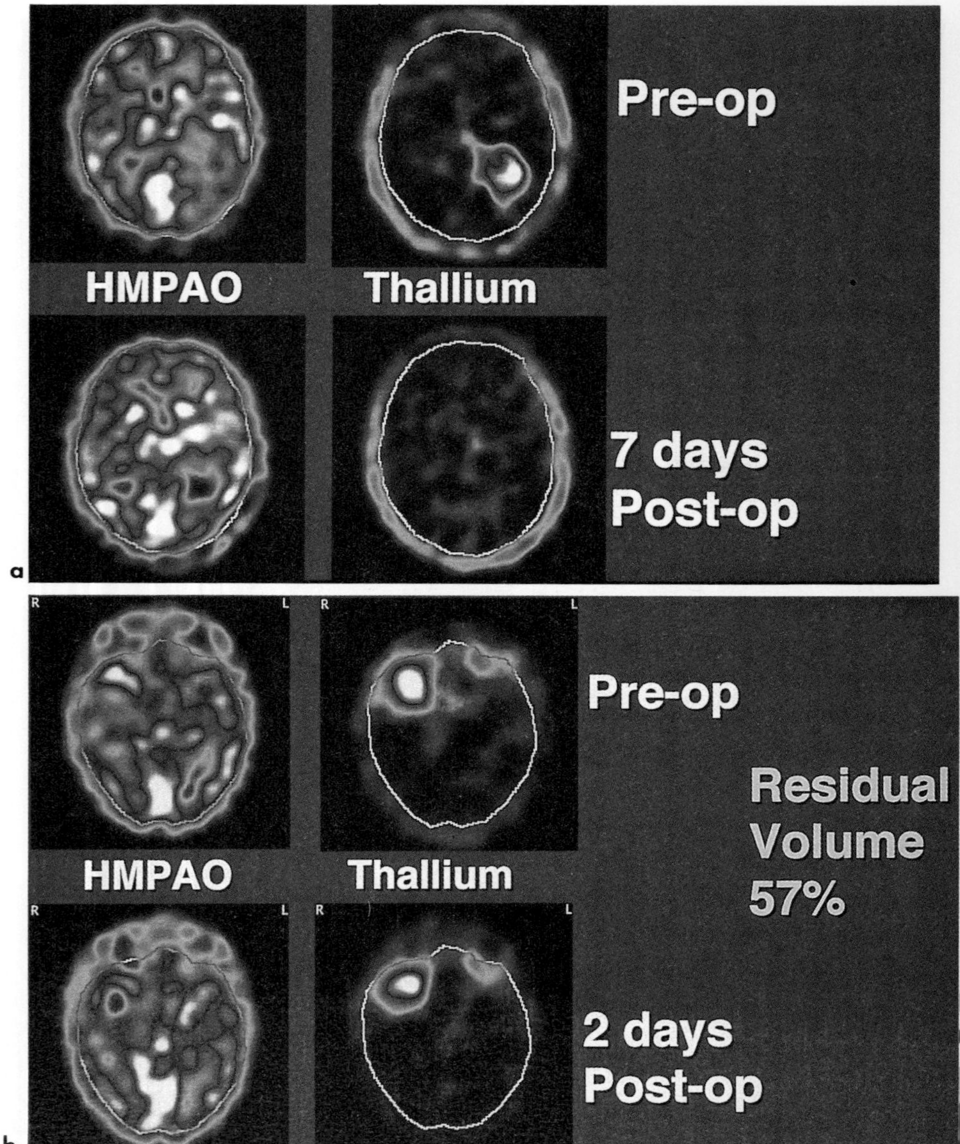

Figure 6.5. (a) Standard pre- (above) and post- (below) operative HMPAO (left) and thallium (right) SPECT images showing complete excision of the thallium positive tissue. (b) Equivalent images for an incompletely excised tumour. (For colour version of this figure see page 187.)

in defining individual prognosis. Albert et al. have shown that descriptions of tumour debulking given by the surgeon can be unreliable[5]. In contrast, where tumour excision has been estimated as complete using early post-operative MRI there has been an increase in survival of up to 20%. If SPECT can improve the

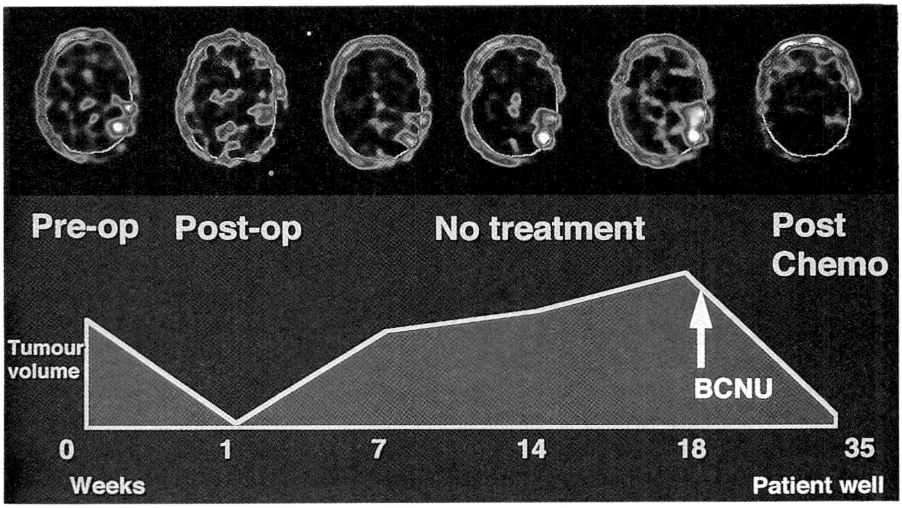

Figure 6.6. How SPECT/thallium uptake can be used to monitor the response of the patient to treatment. A series of scans in a patient with a recurrent oligoastrocytoma who had an excellent response to BCNU. (For colour version of this figure see page 188.)

quality of viable tumour imaging then we might expect a study of survival after SPECT-directed surgery to show comparable, if not better, survival results.

Defining the tissue volume to be treated by radiotherapy: When considering radiotherapy, whether pre- or post-surgical, or at the time of tumour recurrence, the radiotherapist asks the question "What is the target, and what volume shall I treat?". Before surgical interference, the volume of tumour from conventional imaging plus a suitable margin would seem an appropriate base for defining a target. After surgery, or with tumour recurrence, the target shape, volume, and site may become difficult to define, leading not only to larger treatment volumes but possibly to misdirected treatments. The more accurate localization of tumour volume offered by thallium SPECT can allow treatment volume to be reduced.

Monitoring treatment: The ability of SPECT imaging to monitor the effects of therapy is currently under investigation. Figure 6.6 (see also colour section) shows a series of scans in one patient who had an excellent response to BCNU (a cytotoxic nitrosourea). Figure 6.7 (see also colour section) shows the more usual response of a primary tumour to radiotherapy. The ability to detect tumour recurrence before neurological deterioration offers many possibilities for improving and measuring the effect of treatment modalities.

Determining radiation necrosis: Following standard external beam radiation therapy, patients can suffer brain swelling, oedema and non-specific post-radiation symptoms. Radiation necrosis presents with neurological deterioration

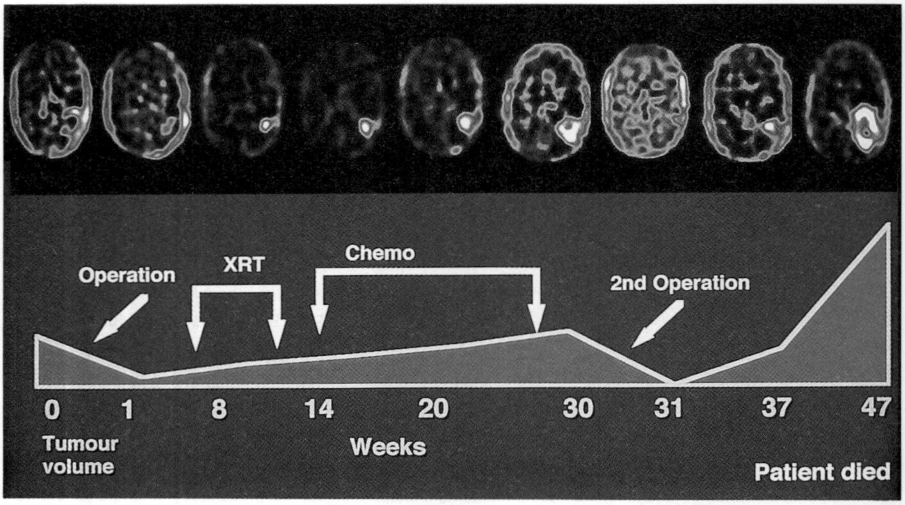

Figure 6.7. Typical patient with a malignant glioblastoma showing the response of the primary tumour to radiotherapy given from week 31 onward. Note the regrowth of the tumour. (For colour version of this figure see page 188.)

weeks to months after completion of the radiation course, in a period when tumour recurrence might also occur. Distinguishing these pathologies by standard CT or MRI imaging is difficult, and a further biopsy may be needed to verify the diagnosis. SPECT has been shown to be very useful in this context. Schwartz et al. concluded that where thallium levels were high, there was recurrent tumour[19]. Where thallium levels were low, increased uptake of HMPAO indicated necrosis. In our experience of these clinical situations, a positive thallium scan in a patient with a proven malignant tumour always indicates recurrence, an experience verified by others[29]. This information provides a valuable basis on which, not only to advise the patient, but also to consider a more rational management programme.

Limitations of thallium SPECT

Eighty percent of malignant brain tumour recurrences occur within the radiation portals, but data from postmortem studies and core sampling have shown malignant cells distant from the lesion. More accurately guided surgery should improve survival, but survival will then be determined by recurrence due to more diffuse disease. Thallium SPECT will detect viable focal concentrations of tumour, but if individual cells within areas of diffuse disease are malignant and take up thallium, their density may well be too low for thallium concentration to achieve detectable levels. The imaging resolution of SPECT systems for thallium is approximately 2–5 ml tumour volume: this figure only has biological relevance where there is no heterogeneity of distribution within the spatial resolution of the imaging system. Areas where malignant cells are present but in relatively low density may have an intact blood–brain barrier, again possibly limiting thallium uptake[36].

These limiting factors can be measured but it is clear that cell density will ultimately fall below the capacity of thallium SPECT visualization, and for that matter possibly any SPECT isotope system. Nonetheless, the conclusion from a recent editorial in the *Journal of Nuclear Medicine*[22], concerning the use of thallium, seems pertinent: "Thallium[201] SPECT now appears to be the method of choice for grading gliomas, assessing residual viable glioma tumours following radiation therapy, and detecting malignant degeneration in low grade gliomas".

Other agents used for imaging brain tumours with SPECT

Idomethyl tyrosine (IMT)

Tyrosine is an amino acid which is non-essential in the presence of phenylalanine, and is involved in both protein synthesis and transmitter production in the brain.

Uptake of amino acids by astrocytes is by the so-called 'L' and 'A' systems which have differential selectivity for the amino acids. Both systems appear to be expressed by both brain and tumour cells, with the A transporter more highly expressed in transformed or malignant cells. The use of amino acids with a preference for the A system offers a method of differentiating viable tumour cells from brain. Blasberg et al. have shown a more than 10-fold difference in specific A uptake activity in C6 glioma cell, over cortex and medulla, in rats[38].

Tyrosine is a freely available amino acid readily released into the plasma after ingestion of food. Its uptake into brain is a specific transporter-mediated process which can be demonstrated by loading the plasma with a competitive amino acid feed and comparing uptake of tyrosine into the brain in loaded and fasted patients. The decrement in tyrosine uptake into normal brain with an amino acid load is of the order of 50%. Some lesions, such as intracerebral metastases and meningiomas, show no decrement, possibly because the transporter system in these lesions is not saturated at the physiological infusion levels used, while the transporter systems of brain and astrocytes is. Further characterization of these systems may allow the development of more selective markers for evaluation of tumour behaviour[39].

Initial studies confirm that IMT is useful as a SPECT agent. Using IMT it has been possible to distinguish not only high grade glial tumours, but also lower grade intrinsic tumours[40]. Langen et al. clearly demonstrated IMT uptake in a brain lesion that was contrast negative, i.e. with an intact blood–brain barrier. The lesion proved to be a grade 2 astrocytoma on biopsy. Such evidence suggests that tyrosine might be a useful marker for low grade lesions, where thallium seems to be less sensitive. More systematic evidence is required. It has yet to be determined whether definitive treatment of low grade lesions can protect against the onset of degenerative malignancy, and IMT SPECT might allow an objective assessment of the results of treatment.

There are a number of practical difficulties with the use of IMT as a tumour marker. The radiolabelling of tyrosine can often give a barely acceptable yield, and must be prepared 12h or so before use. This makes it very expensive as an imaging agent. The uptake of IMT into cells is slow, is related to metabolic activity, and has a variable washout period (at present, most authors measure uptake at approximately 10–15 min post-injection, aiming for a peak level). There is often spurious uptake in the cerebellum that can render tumour/cerebellar ratios invalid (unpublished personal data).

There are, however, a number of advantages in using a marker whose mode of accumulation in cells is different from that of thallium. IMT uptake is independent of damage to the blood–brain barrier, and hence may be more suitable for looking at lower grade tumours where the blood–brain barrier remains more or less intact.

^{99}Technetium methoxybutyl isonitrile (MIBI)

^{99}Technetium methoxybutyl isonitrile (MIBI) has been reported to show results comparable to ^{201}Tl for the evaluation of childhood brain tumours and of recurrent tumours[31]. Relative uptake indices appear superior, but high uptake in the choroid plexus poses problems with image interpretation.

Iododeoxyuridine (IUDR)

Iododeoxyuridine has been used to image tumour proliferation and, as a nucleotide base analogue, might be expected to be useful as a prognostic marker[11,12]. Studies that have used a brominated analogue which is incorporated into DNA during cell proliferation have shown variably low labelling index in glioblastoma biopsies. One would expect such aggressive tumours to demonstrate a high index, and indeed this has been seen in cell culture. The explanation of this discrepancy is unknown, but may be that IUDR is only taken up in vivo only by a small pool of dividing cells. Its potential use in the intrinsic brain tumours seems limited, but may have a niche in the evaluation of some metastases or possibly cerebral lymphoma.

SPECT methods using combinations of tracers

It is generally thought that glioblastomas often arise from low grade astrocytic tumours that undergo malignant degeneration: most evidence suggests that degenerative malignancy occurs in 50% of low grade tumours over a 5-year period. Such a model would suggest that glioblastoma consists partly of a focal malignant portion, able to be imaged by tracers with specific uptake and associated with blood–brain barrier disruption. In addition, there would be more diffuse lower grade tumour surrounding the focus of malignant tumour. Tumour

Table 6.1. Three glioma cell types can be defined according to their patterns of thallium/tyrosine uptake.

Tumour cell type	Thallium uptake	Tyrosine uptake
A	+	+
B	–	+
C	–	–

recurrence would be determined not only by residual malignant tumour, but by adjacent low grade tumour with malignant potential. This lower grade tumour field would have an intact blood–brain barrier, and a low affinity for thallium, so combination studies using thallium and tyrosine might offer the possibility of defining these different tumour compartments, compartments which might demand radically different treatment strategies.

Our studies in a limited number of patients have shed some light on this theory. In some cases, while thallium uptake has been isotopic with gadolinium uptake, tyrosine SPECT has shown significant uptake over a larger area, isotopic with the enhancing+non-enhancing (MRI) area of the tumour. This suggests that these imaging techniques can sometimes identify two cell types. Low grade tumours (high T2 signal, no gadolinium enhancement, intact blood–brain barrier) should show tyrosine uptake but no thallium uptake but, in our experience, some such low grade lesions show neither thallium nor tyrosine uptake. Thus, there exists the possibility of at least three astrocytic cell types defined by their combination of thallium and tyrosine uptake (Table 6.1).

Nitromidazole and liposomes

As noted in the introduction, glioblastomas appear more resistant to radiation in vivo than in vitro. It has been proposed that this is due to hypoxia within tumours, for which evidence exists[41]. Chapman and others have tried to image hypoxia in brain tumours using a novel nitromidazole compound[42], but were unsuccessful in demonstrating it, possibly because of the microenvironmental area of the hypoxia.

A more interesting and novel way of using SPECT to monitor treatment modalities has been demonstrated by Rampling et al.[43]. Using radiolabelled liposomes, they have been able to demonstrate targeting of liposomes to brain tumours in such a way that access of drugs to the tumour can be quantified. This technique offers possibilities for classifying tumours and assessing their suitability for different treatment protocols.

Conclusions

Any new clinical technique must be judged finally by its ability to improve the care of patients. Thallium SPECT aids tumour diagnosis, facilitates accurate surgical treatment and radiotherapy, and offers the possibility of objective treatment monitoring. With the addition of such agents as tyrosine, the added possibility of managing low grade tumours more rationally can be pursued.

Integration of SPECT with other imaging modalities and clinical data offers a finer basis for informed decision making in the management of intrinsic brain tumours. This should assist clinicians in evolving a rational management of these most difficult of malignant conditions.

References

1. Kim TS, Halliday AL, Hedley-White ET, Convery K. Correlates of survival and the Daumas–Duport grading system for astrocytomas. J Neurosurg. 1991;74:27–37.
2. Chang CH. Comparison of post operative radiotherapy and combined post operative radiotherapy and chemotherapy in the multidisciplinary management of malignant glioma. Cancer. 1983;52:937–1007.
3. Bloom HJG. Intracranial tumors: response and resistance to therapeutic endeavours 1970–1980. Int J Radiat Oncol Biol Phys. 1982;8:1083–113.
4. Stenning SP, Freedman LS, Bleehan NM. An overview of published results from randomised studies of nitrosourease in primary high grade malignant gliomas. Br J Cancer. 1987;56:89–90.
5. Albert FK, Forsting M, Sartor K, Adams H-P, Kunze S. Early post-operative magnetic resonance imaging after resection of malignant glioma objective evaluation of residual tumour and its influence on regrowth and prognosis. Neurosurg. 1994;34:445–61.
6. Hochberg F, Pruitt A. Assumptions in the radiotherapy of glioma. Neurology. 1980;30:997–9.
7. Fertil B, Malaise E. Intrinsic radiosensitivity of human cell lines is correlated with radiosensitivity analysis of 101 published survival curves. Int J Radiat Oncol Biol Phys. 1985; 11:699–707.
8. Steel GG, Peacock JH. Why are some human tumors more radiosensitive than others? Radiother Oncol. 1989;15:63–72.
9. Tahian A, Budach J, Freeman D, Giono H, Smit H, Steele E. In vivo radiation sensitivity of glioblastoma multiforme. J Neuro-oncol. 1994;19:179.
10. Taghan A, Ramsay J, Allalunis-Turner J, et al. Intrinsic radiation sensitivity may not be the major determinant of the poor clinical outcome of glioblastoma multiforme. Int J Radiat Oncol Biol Phys. 1993;25:243–9.
11. Hoshino T, Barker M, Wilson L. Cell kinetics of human gliomas. J Neurosurg. 1972;37:15–26.
12. Nagashima TS, DeArmond S, Murcovic J. Immunocytochemical demonstration of S-phase cells by anti-bromodeoxuridine monoclonal antibody in human brain tumour tissues. Acta Neuropathol. 1985;67:155–9.
13. Atkins HL, Budinger TE, Labowitz E. Thallium201 for medical use. Part 3: Human distribution of physical imaging properties. J Nucl Med. 1977;18:133–40.
14. Ancri D, Bassett JY, Lonchampt MF, Etarard C. Diagnosis of cerebral lesions by thallium201. J Nucl Med. 1978;128:417–22.
15. Kaplan WD, Takronan T, Morris JH, Rumbough CL, Connolly BT, Atkins HL. Thallium201 brain tumor imaging: a comparative study with pathologic correlation. J Nucl Med. 1987;28: 47–52.
16. Kim KT, Black KL, Marciano D. Thallium^{201} SPECT imaging of brain tumors: methods and results. J Neurosurg. 1989;71:342–6.
17. Ueda T, Kaji Y, Wakisaka S. Time sequential single photon emission computed tomography studies in brain tumour using thallium-201. Eur J Nucl Med. 1993;20:138–45.
18. Jinnouchi S, Hoshi H, Ohnishi T, et al. Thallium-201 SPECT for predicting histological types of meningiomas. J Nucl Med. 1993;34:2091–4.
19. Schwartz RB, Carvalho PA, Alexander E, Loeffler JS, Folkerth R, Holman BL. Radiation

necrosis vs high-grade recurrent glioma: Differentiation by using dual-isotope SPECT with 201Tl and 99m Tc-HMPAO. Am J Neuroradiol. 1991;12:1187–92.
20. Kubota R, Kubota K, Yamada S. Methionine uptake by tumor tissue: a microautoradiographic comparison with FDG. J Nucl Med. 1995;36:484–92.
21. Britton J, Blank M. Thallium activation of the (Na+K+) activated ATPase of rabbit kidney. Biochem Biophys Acta. 1968;159:160–6.
22. Tonami N, Hisada K. Thallium201 SPECT in the evaluation of gliomas. J Nucl Med. 1993; 34:2089–90.
23. Whittle I. The biology of glioma. In: Teasdale GM, Miller JD, editors. Current Neurosurgery. Churchill Livingstone, 1992:255–84.
24. Lopez M, van Den Berg S, Scheithauer B. The World Health Organisation classification system of nervous system tumours in experimental neuro-oncology. In: Levine AJ, Schmidek HH, editors. Molecular genetics of nervous system tumours. New York: Wiley Liss, 1993:1–36.
25. Black KL, Hawkins RA, Kim KT, Becker DP, Lerner C, Marciano D. Use of Thallium201 (SPECT) to quantitate malignancy grade of glioma. J Neurosurg. 1989;71:342–6.
26. Francavilla T, Miletich R, Di Chiro G. Positron Emission Tomography in the detection of malignant degeneration of low-grade gliomas. Neurosurg. 1989;24:1–5.
27. Mountz JM, Stafford-Schuck M, McIver PE. Thallium^{201} tumour/cardiac ratio estimation of residual astrocytoma. J Neurosurg. 1988;68:705–9.
28. Ancri D, Bassett JY. Diagnosis of cerebral metastases by Thallium201. Br J Radiol. 1980; 53:443–5.
29. Vertosick FT, Selker RG, Grossman SJ, Joyse JM. Correlation of Thallium201 single photon emission computed tomography and survival after treatment failure in patients with glioblastoma multiforme. Neurosurg. 1994;34(3):396–401.
30. Piepmeier JM. Observations on the current treatment of low grade astrocytic tumors of the cerebral hemispheres. J Neurosurg. 1987;67:177–81.
31. O'Tuama LA, Treves ST, Larar JN, Packard AB, Kwan AJ, Barnes PD. Thallium201 versus Technetium-99m-MIBI SPECT in evaluation of childhood brain tumors: A within-subject comparison. J Nucl Med. 1993;34:1045–51.
32. Levivier M, Goldman S, Pirotte B, et al. Diagnostic yield of stereotactic brain biopsy guided by positron emission tomography with [18F]Fluorodeoxyglucose. J Neurosurg. 1995;82: 445–52.
33. Mosskin M, Ericson K. Positron Emission Tomography compared with magnetic resonance imaging and computed tomography in supratentorial gliomas using multiple stereotactic biopsies as reference. Acta Radiologica. 1989;30:225–32.
34. Maciunas RJ, Kessler RM, Maurer C, Mandava V, Watt G. Positron emission tomography image directed stereotactic neurosurgery. Stereotact Funct Neurosurg. 1992;58:134–40.
35. Hughes SW, Sofat A, Kitchen ND, et al. Computer planning of stereotactic iodine-125 seed brachytherapy for recurrent malignant gliomas. Br J Radiol. 1995;68:175–81.
36. Drake J, Rutka J, Hoffman H. ISG Viewing Wand System. Neurosurg. 1994;34:1094–5.
37. Kelly D, Hunt C. The limited value of cytoreductive surgery in elderly patients with malignant gliomas. Neurosurg. 1994;34:62–7.
38. Tjuvajev KG, Abrams DJ, Finn RG, Blasberg RG. Imaging brain tumours with non-metabolised amino acids: advantages over fluorodeoxyglucose (FDG). J Cereb Blood Flow Metab. 1993;13:S572.
39. Oldendorf WH. Saturation of amino acid uptake by human brain tumour demonstrated by SPECT. J Nucl Med. 1991;32:1229–30.
40. Langen K-J, Goenen HH, Roosen N. SPECT studies of brain tumours with L=3-[123I] Iodo-(-Methyl Tyrosine): Comparison with PET, 124IMT and first clinical results. J Nucl Med. 1990;31:281–6.
41. Cruickshank GS, Rampling RP, Cowans W. Direct measurement of the pO2 distribution in human malignant brain tumours. Adv Exp Med Biol. 1993;345:465–70.
42. Groshar D, McEwan JB, Parliament MB, et al. Imaging tumour hypoxia and tumour perfusion. J Nucl Med. 1993;34:885–8.

43. Khalifa A, Rampling R, Patterson J, Murray T, Kerr D. Investigation of the uptake and distribution of Irridium111 labelled liposomes in relapsed high grade astrocytoma. JNNP. 1995;59:204.

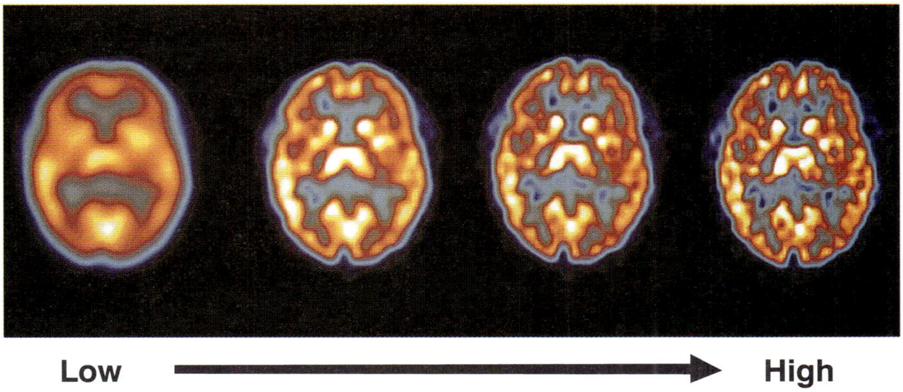

Figure 1.9. How the same raw data (in this case that making up the single illustrated orbitomeatal slice) may be filtered to different degrees to give different levels of resolution and 'noise' in the final image.

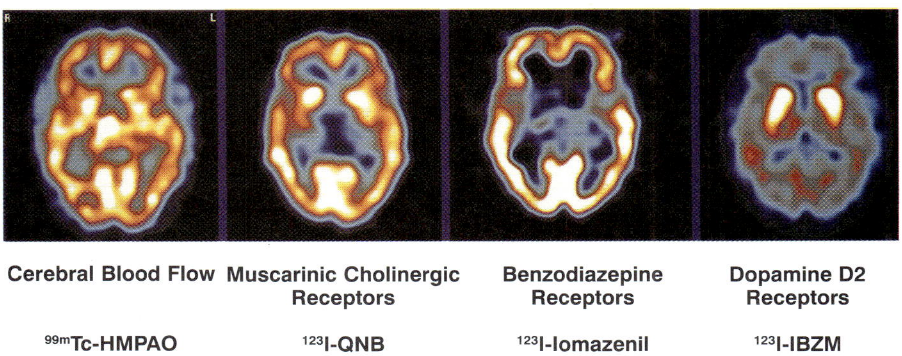

Figure 1.12. SPECT images illustrating different uptake patterns obtained using different radionuclides.

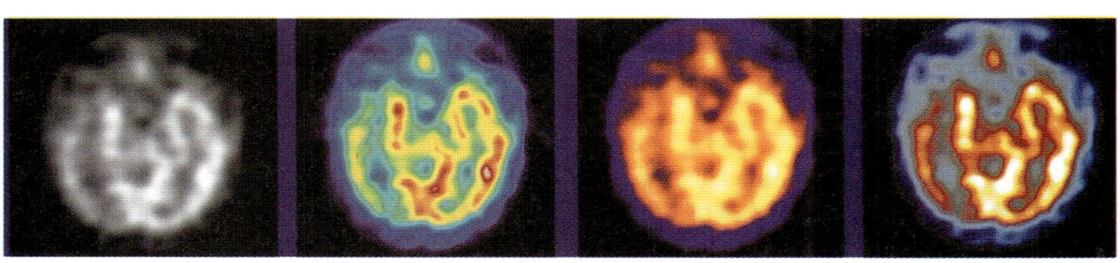

Figure 1.13. The same slice is displayed using four colour scales (left to right). In all cases, no signal is black. (a) Monochrome grey scale, with white showing the highest signal, black the lowest. (b) 'Rainbow' type colour scale using a combination of colours, rising through blue, green, yellow, red and white (only a very small area of white is seen in the posterolateral left temporal cortex. (c) Monochrome 'hot metal' type red scale, from dark red to light yellow/white. (d) Bichrome blue/red 'fire and water' colour scale, with reds, yellows and whites showing the higher signal levels.

R. Duncan (ed): SPECT Imaging of the Brain, 179–188.

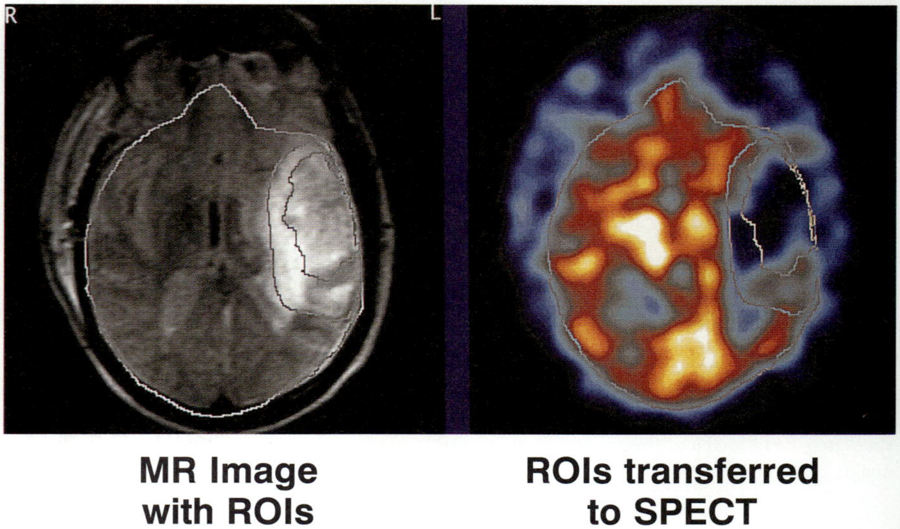

Figure 1.15. Placement of regions of interest using anatomical image data. The MR and SPECT images are co-registered. ROIs are drawn around anatomical features of interest on the MR images, and the corresponding SPECT slice is then redisplayed with the MR-determined ROI in place.

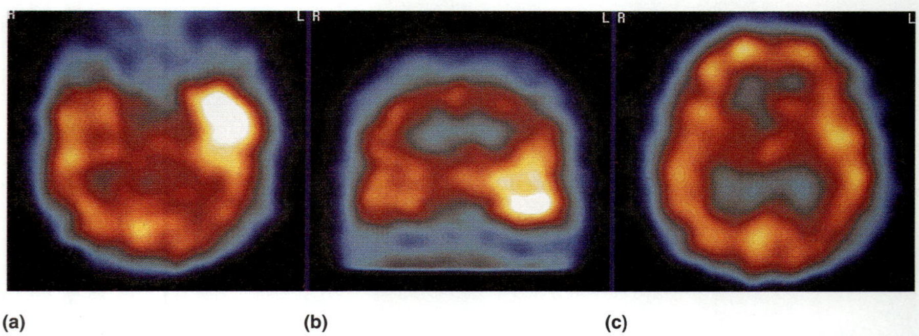

Figure 2.8. Ictal HMPAO SPECT injection during a complex partial seizure of left mesial temporal origin. All three images are from the same dataset. (a) Axial slice in the long axis of the temporal lobe showing marked hyperperfusion of the whole left temporal lobe. (b) Coronal slice through the mid-temporal lobe, showing hyperperfusion of the whole left temporal lobe, extending into the adjacent frontal cortex. More superiorly, there is mild hypoperfusion on the left frontal cortex. (c) Axial slice in the orbitomeatal plane, showing hyperperfusion of left basal ganglia and thalamus.

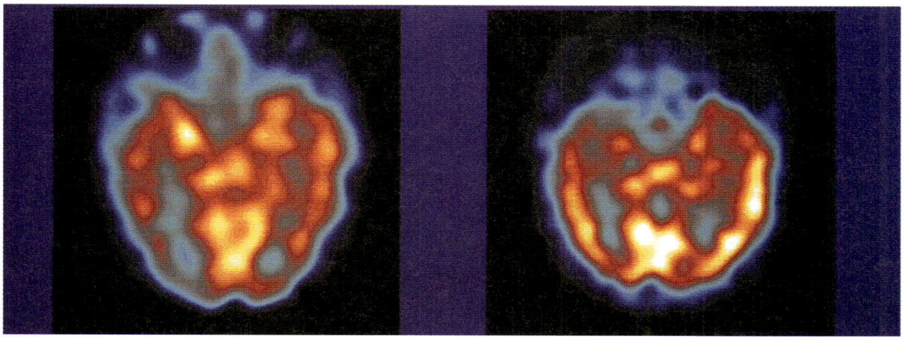

(a) (b)

Figure 2.9. (a) HMPAO injection 1 min after the end of a complex partial seizure of right mesial temporal origin. Axial slices in the long axis of the temporal lobe, showing hypoperfusion of the right lateral temporal cortex. This extended to involve the whole of the right hemisphere, with the exception of the right mesial temporal cortex, where blood flow remains high. (b) Right mesial temporal seizure in a different patient, in whom HMPAO was injected 6 min after the end of a right mesial temporal lobe seizure. This shows hypoperfusion of both mesial and lateral temporal cortex on the right.

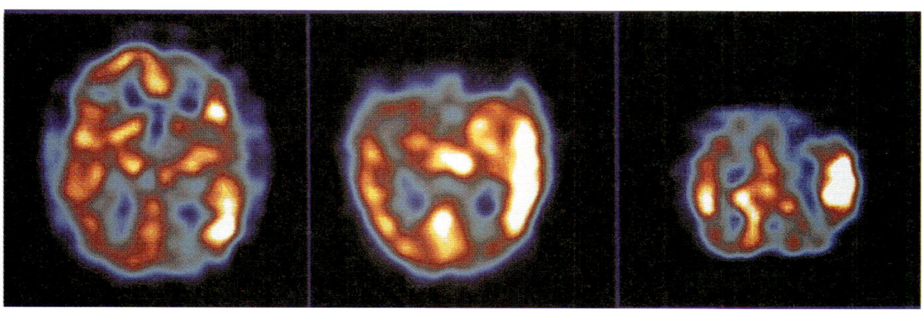

(a) (b) (c)

Figure 2.10. HMPAO SPECT. Injection during a complex partial seizure originating in area of the left temporo-parieto-occipital junction (TPOJ). All three images are from the same dataset. (a) Axial slice showing intense hyperperfusion in the area of the TPOJ, with a small focus of hyperperfusion in the ipsilateral inferolateral frontal cortex, with hypoperfusion of the rest of the ipsilateral frontal lobe. (b) Axial slice in the long axis of the temporal lobe showing hyperperfusion of the whole temporal lobe. Note that, unlike the pattern seen in mesial temporal lobe seizures, hyperperfusion is clearly more marked in the lateral as opposed to the mesial temporal cortex. (c) Posterior coronal slice, showing hyperperfusion of the TPOJ, with intense surrounding hypoperfusion. There is a small area of hyperperfusion in the contralateral parietal cortex.

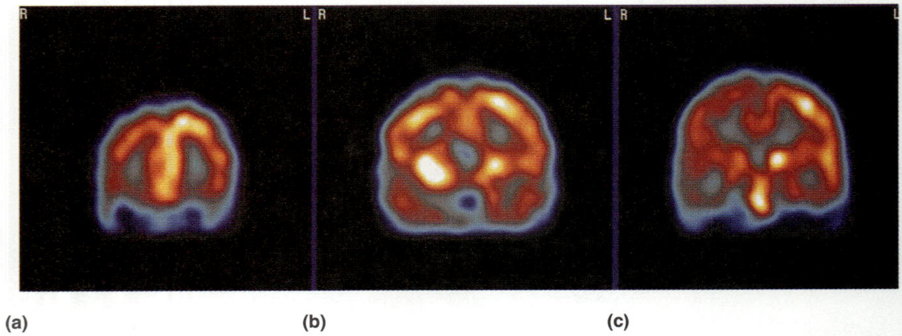

Figure 2.11. HMPAO injection during a complex partial seizure of left mesial frontal origin (MRI and pathological examination indicated focal cortical dysplasia at this site). (a) Coronal slice through the anterior frontal lobe, showing left mesial frontal hyperperfusion. (b) Coronal slice through the basal ganglia, frontal and temporal lobes, showing hyperperfusion of the right basal ganglia, with relative hypoperfusion of the temporal lobes. (c) More posterior coronal slice showing hyperperfusion of the left basal ganglia extending down into the brain stem. The posterior part of the right cortical mantle is hypoperfused.

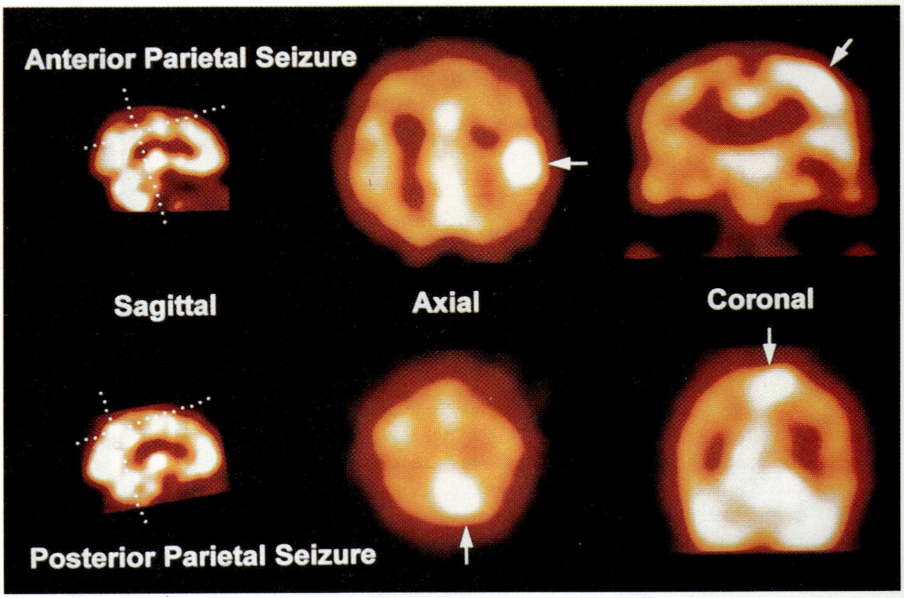

Figure 2.12. Parietal lobe epilepsy. The upper images show a seizure arising from the anterior parietal lobe in a patient whose seizure comprised tingling in the right hand followed by asymmetrical bilateral dystonic posturing. There is hyperperfusion seen in the anterior parietal area. The lower images show a posterior parietal seizure, manifested by an experiential aura followed by a psycho-paretic complex partial seizure. The hyperperfusion is more posterior. (Illustration kindly supplied by Dr S. Berkovic, Austin Hospital, Melbourne, Australia).

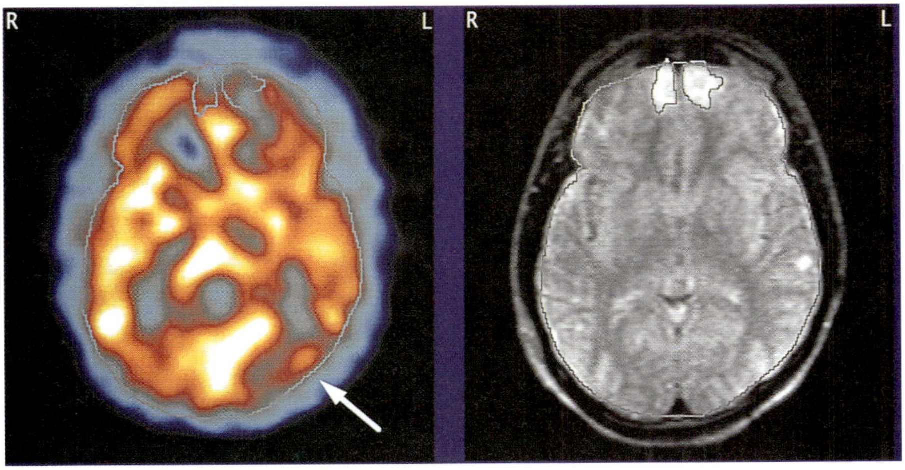

(a) (b)

Figure 3.1 Images from a 23-year-old woman who fell from a loft. She was conscious on day 5, but very uncooperative and aggressive. (a) T2 weighted MRI shows bifrontal lesions (outlined). (b) On the SPECT image there are matching frontal flow deficits; there is also a 'contrecoup' left occipital perfusion defect with no matching structural lesion (arrow).

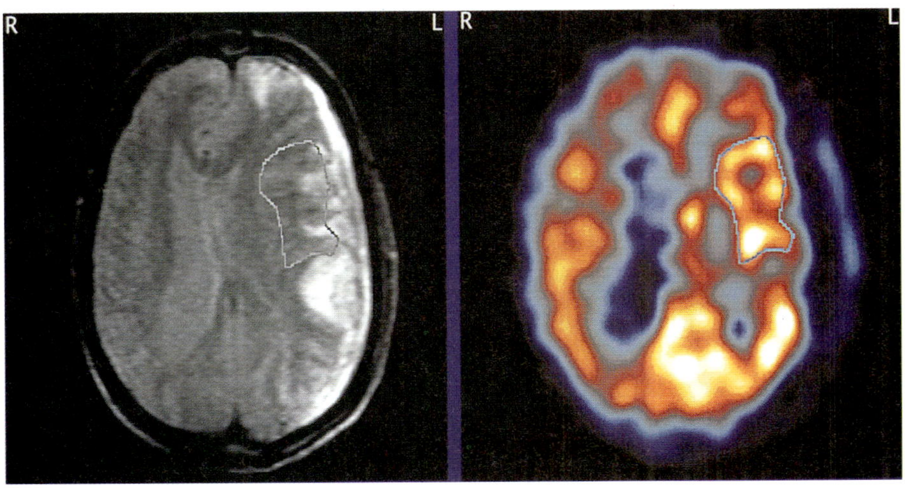

Figure 3.2 A patient with head injury of moderate severity complicated by a left subdural haematoma. MRI and HMPAO SPECT were performed 5 days following injury. The MRI image shows the subdural collection, with an underlying contusion, seen as an area of increased T2 signal just inferior to the marked region of interest. The corresponding SPECT image shows the physical effects of the haematoma and a small area of hypoperfusion in association with the contusion. There is, however, marked hyperperfusion anterior to the contusion which falls within MRI-normal tissue (as marked by the ROI placed on both co-registered images), and a similar area posterior to the contusion.

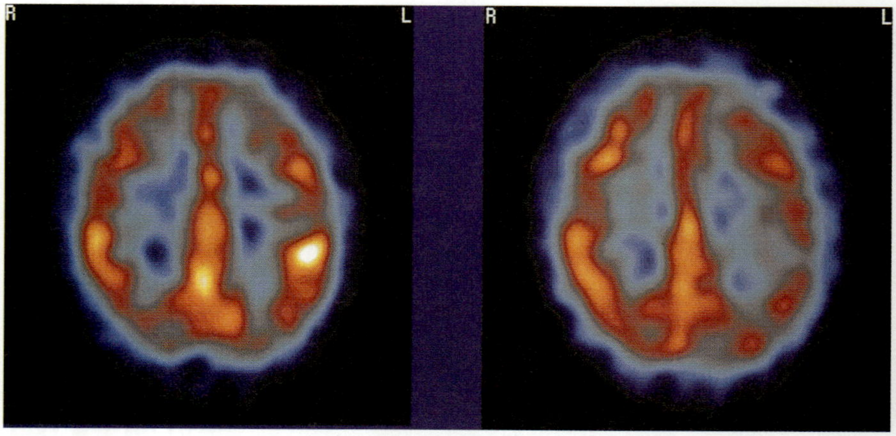

Figure 4.4. A 30-year-old man suffered a parieto-occipital stroke, with visual field defect. This high axial slice shows luxury perfusion at the anatomic site of the lesion at 7 days, which has resolved at 6 months.

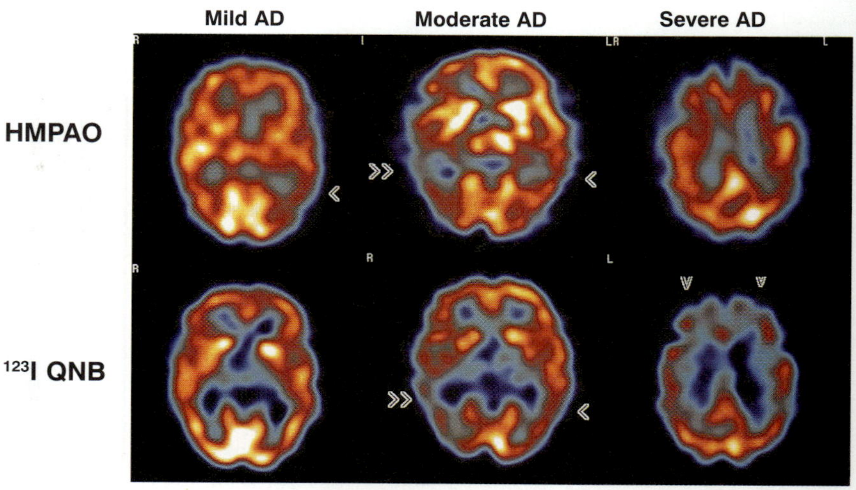

Figure 5.5. SPECT scans of the distribution of [^{123}I]QNB at 21 h compared with [^{99}Tc^{m}]HMPAO in patients with clinically diagnosed Alzheimer's disease and varying severity of cognitive impairment. The top scans are [^{99}Tc^{m}]HMPAO and the lower ones [^{123}I]QNB. The patients are: (a) a mildly demented AD patient with CAMCOG of 62; (b) a moderately demented patient with CAMCOG of 32; (c) a severely demented patients who was given a CAMCOG score of zero as she was untestable. Relative to [^{99}Tc^{m}]HMPAO (which gives a measure of general regional function as determined by blood flow), the focal deficits in the [^{123}I]QNB pattern at 21 h (which is governed by muscarinic receptor density) appear to be less marked in the mild case and more marked in the severe case. This suggests that muscarinic receptors are not selectively affected until late in the disease process.

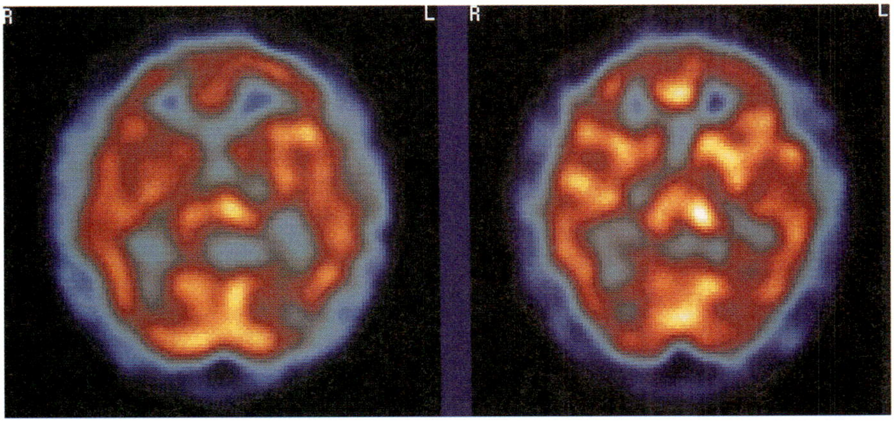

(a) (b)

Figure 5.8. SPECT scans with [$^{99}Tc^m$]HMPAO at the level of the basal ganglia from (a) a drug-naive schizophrenic patient and (b) the same patient on neuroleptics, showing an increase in blood flow in the basal ganglia.

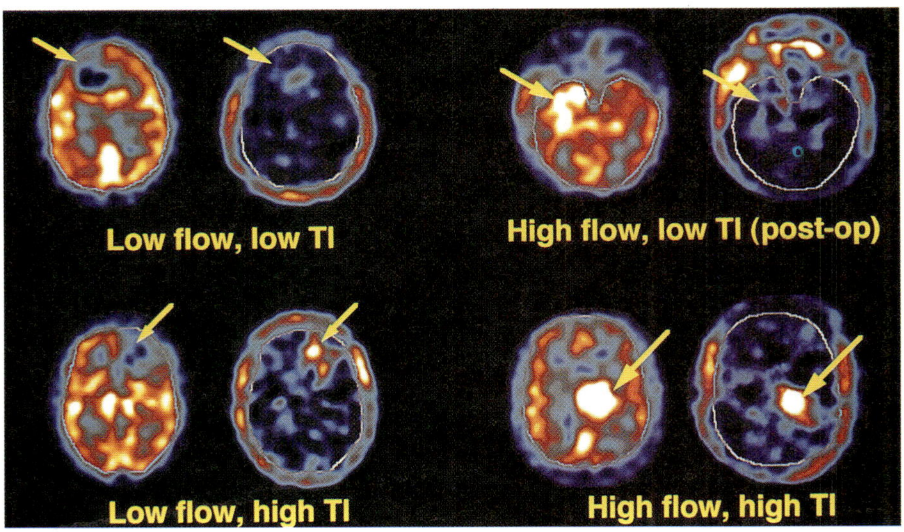

Figure 6.2. Four combinations of blood flow (demonstrated by HMPAO) and thallium uptake by tumours. Of each pair of images, the HMPAO image is on the left, thallium on the right. There does not appear any clear cut relationship between the two.

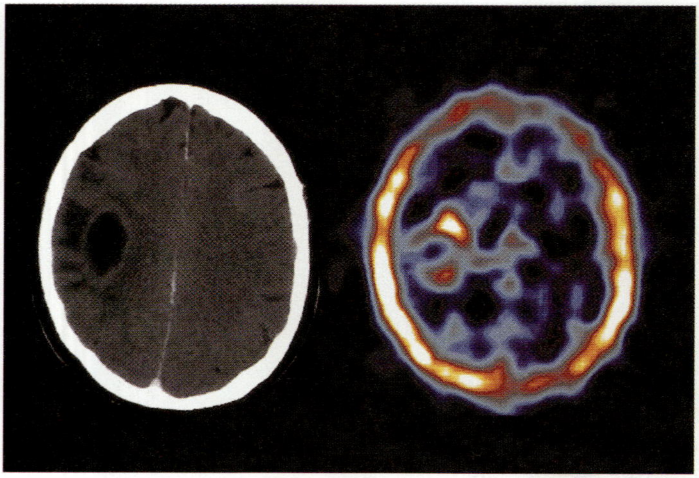

CT+contrast **Tl-201**

Figure 6.3. CT scan showing a high grade cystic lesion in the right parietal lobe. The simple approach for biopsy would suggest biopsy of the outer wall of the cyst. The SPECT scan shows, however, that the area of maximum thallium activity is in the anterior medial wall of the cyst, and that this is in fact the most sensible target for biopsy.

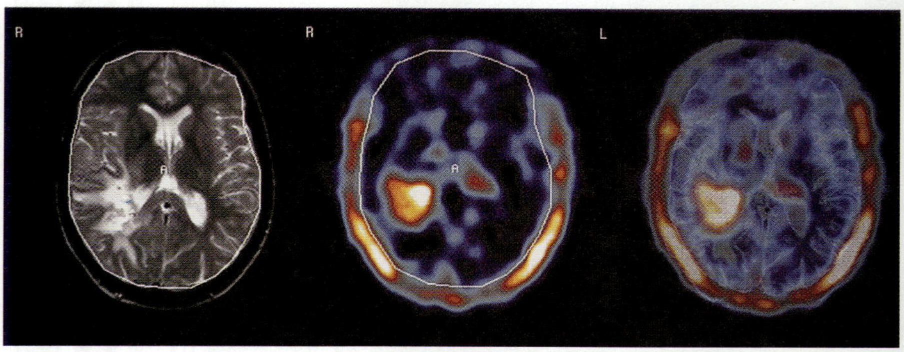

T2 weighted MRI **T1-201** **Overlay**

Figure 6.4. MRI scan (left) shows a contrast-enhanced tumour in the left trigonal area. SPECT image (right) co-registered with the MRI so that the limits of the proliferating tumour volume can be determined in terms of MRI stereotaxic coordinates.

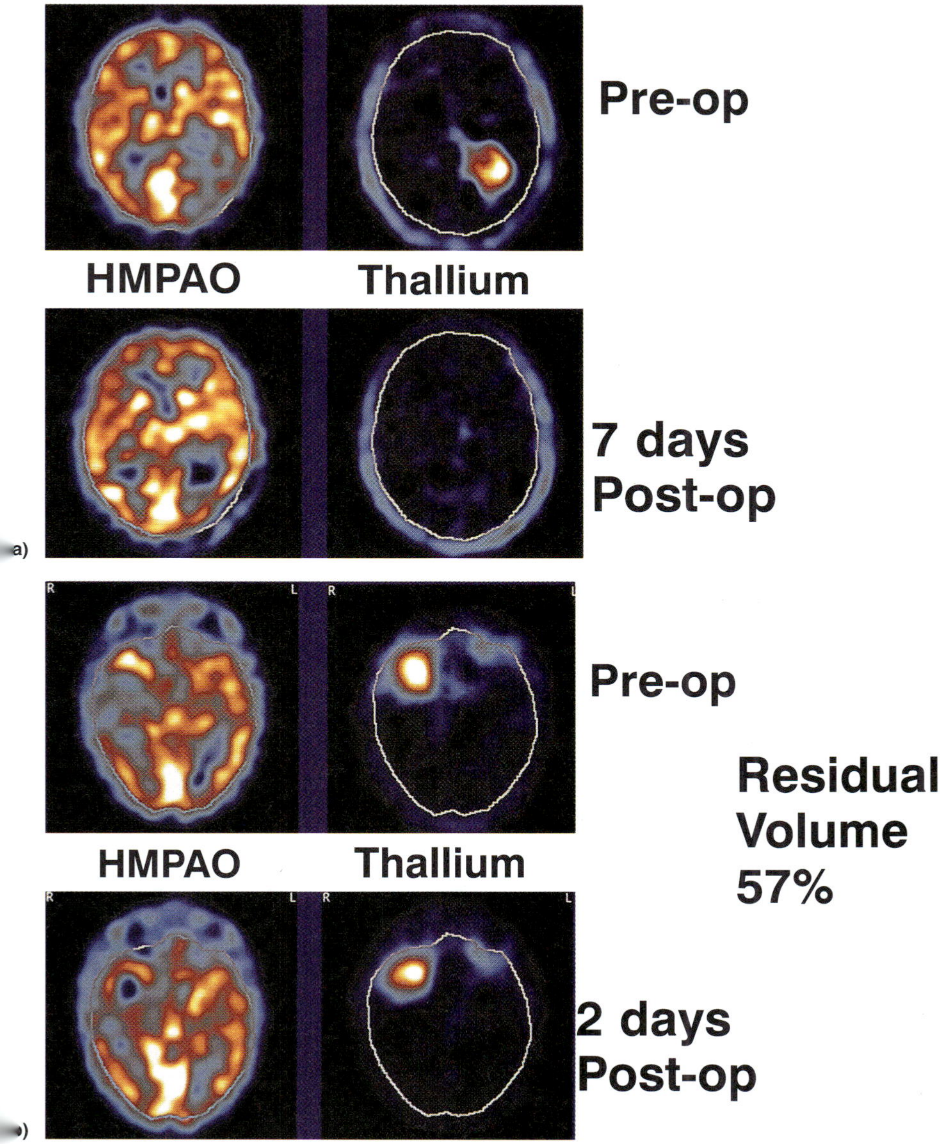

Figure 6.5. (a) Standard pre- (above) and post- (below) operative HMPAO (left) and thallium (right) SPECT images showing complete excision of the thallium positive tissue. (b) Equivalent images for an incompletely excised tumour.

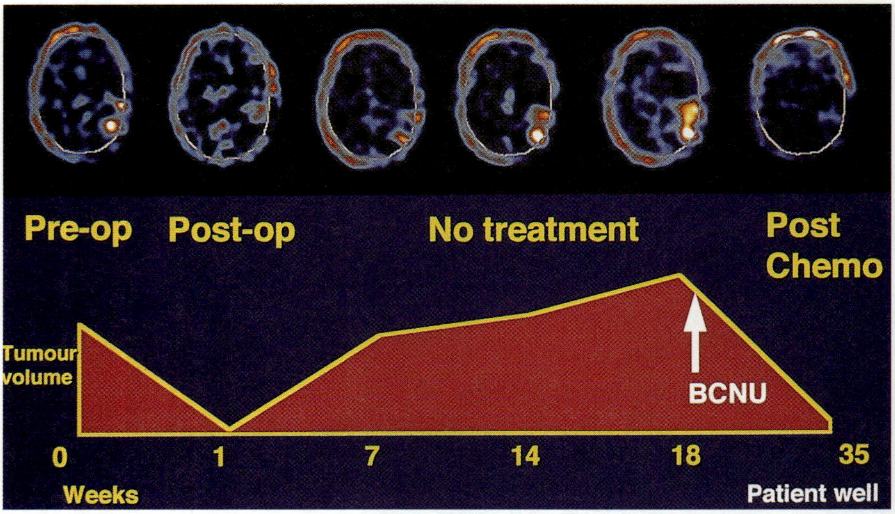

Figure 6.6. How SPECT/thallium uptake can be used to monitor the response of the patient to treatment. A series of scans in a patient with a recurrent oligoastrocytoma who had an excellent response to BCNU.

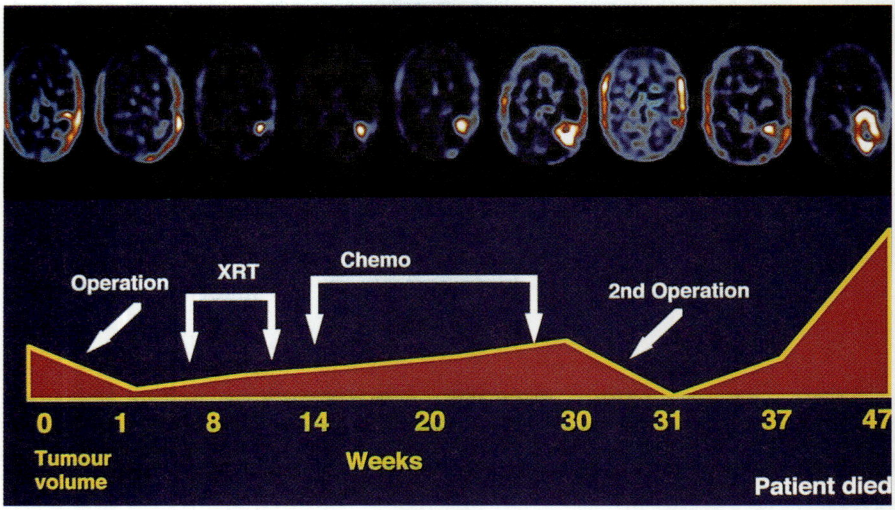

Figure 6.7. Typical patient with a malignant glioblastoma showing the response of the primary tumour to radiotherapy given from week 31 onward. Note the regrowth of the tumour.

Index

R. Duncan (ed): SPECT Imaging of the Brain, 189–193.

Developments in Nuclear Medicine

20. J.C. Baron, D. Comar, L. Farde, J.L. Martinot and B. Mazoyer (eds.): *Brain Dopaminergic Systems: Imaging with Positron Tomography.* 1991 ISBN 0-7923-1476-X
21. M.K. Dewanjee: *Radioiodination.* Theory, Practice, and Biomedical Application. 1991 ISBN 0-7923-1491-3
22. P.A. Schubiger and G. Westera (eds.): *Progress in Radiopharmacy.* 1992 ISBN 0-7923-1525-1
23. B.M. Mazoyer, W.D. Heiss and D. Comar (eds.): *PET Studies on Amino Acid Metabolism and Protein Synthesis.* 1993 ISBN 0-7923-2076-X
24. G. Stöcklin and V.W. Pike (eds.): *Radiopharmaceuticals for Positron Emission Tomography.* Methodological Aspects. 1993 ISBN 0-7923-2340-8
25. P.H. Cox and M. Pillay (eds.): *The Clinical Applications of SPET.* 1995 ISBN 0-7923-3187-7
26. D. Comar (ed.): *PET for Drug Development and Evaluation.* 1995 ISBN 0-7923-3716-6
27. R.M. Lambrecht: *Biological Models in Radiopharmaceutical Development.* 1996 ISBN 0-7923-3836-7
28. R. Bares and G. Lucignani (eds.): *Clinical PET.* 1996 ISBN 0-7923-4160-0
29. R. Duncan (ed.): *SPECT Imaging of the Brain.* 1996 (forthcoming) ISBN 0-7923-4186-4
30. S.J. Mather (ed.): *Current Directions in Radiopharmaceutical Research and Development.* 1996 ISBN 0-7923-4254-2

Kluwer Academic Publishers - Dordrecht / Boston / London